转型发展系列教材

土木工程实用英语
Practical English for Civil Engineering

刘 娟 李春香 编著

西南交通大学出版社
·成都·

图书在版编目（CIP）数据

土木工程实用英语 = Practical English for Civil Engineering／刘娟，李春香编著． —成都：西南交通大学出版社，2017.7（2019.6重印）
转型发展系列教材
ISBN 978-7-5643-5600-2

Ⅰ. ①土… Ⅱ. ①刘… ②李… Ⅲ. ①土木工程–英语–高等学校–教材 Ⅳ. ①TU

中国版本图书馆 CIP 数据核字（2017）第 169424 号

转型发展系列教材
土木工程实用英语

刘 娟 李春香 编著

责任编辑	姜锡伟
助理编辑	张文越
封面设计	严春艳
出版发行	西南交通大学出版社 （四川省成都市金牛区二环路北一段 111 号 西南交通大学创新大厦 21 楼）
发行部电话	028-87600564　028-87600533
邮政编码	610031
网址	http://www.xnjdcbs.com
印刷	成都中永印务有限责任公司
成品尺寸	185 mm × 260 mm
印张	9.25
字数	253 千
版次	2017 年 7 月第 1 版
印次	2019 年 6 月第 2 次
书号	ISBN 978-7-5643-5600-2
定价	24.00 元

课件咨询电话：028-87600533
图书如有印装质量问题　本社负责退换
版权所有　盗版必究　举报电话：028-87600562

转型发展系列教材编委会

顾　　问：蒋葛夫

主　　任：汪辉武

执行主编：蔡玉波　陈叶梅　贾志永　王　彦

前言
Preface

为了适应社会现实需求，高等教育急需培养应用型人才。土木工程专业英语是土木工程专业教学中非常重要的课程，也是培养应用型人才和现场工程师不可缺少的课程。专业英语作为基础英语的拓展课程，可以扩展学生的专业词汇，培养学生现场交际能力、专业文献的阅读能力和专业文献的翻译能力。本书结合最新的工程技术、新材料的应用发展，引用了大量的工程实例，可作为广大土木工程专业学生以及土木工程技术人员提高专业英语阅读和翻译能力的教材和参考书籍。

本书的对话和课文参考了大量的英文原版书籍，包括了建筑材料、招投标、工程管理、岩土工程、道路、桥梁、高铁等常用的专业词汇。本书取材时考虑难度适中并实用等因素，既重视专业领域相关知识的传递，又注重英语语言现场交际能力的培养；既注意土木工程专业英语材料的阅读，又侧重实际施工场景的对话模拟。课文中还编排了一些与文章相关的插图，使得教材图文并茂，对学生领会课文大有益处。

本教材包括8个单元，主要内容有土木工程概况、建筑材料、施工合同、项目管理、岩土工程、道路工程、桥梁工程、铁路工程。本教材的对话和文章篇幅长短适中，保证2个对话在2学时内完成教学，1篇课文在2学时内完成教学，适用于32学时的教学安排，教师在使用本教材时可根据教学要求和安排灵活把握。为了方便学生课后自学，本教材所有对话、课文的课后练习都有参考答案，课文还有对应的译文。

本教材由西南交通大学希望学院刘娟、李春香编著，本书在编写过程中得到了西南交通大学希望学院唐逸萍、王建、曾文丽、陈诗、陈荟竹、党利、冯丹、王小芳的大力支持，特此表示感谢。全书由李春香进行统稿。

本教材参考并采用了大量的英文文献资料，在此对相关作者表示真心的感谢。

本教材在编写过程中难免存在不足，恳请广大师生给予指正并将意见反馈给我们，邮箱为vanilla@163.com。

编著者
2017年3月

目　录

Unit 1　Civil Engineering ·· 001
Unit 2　Civil Engineering Materials ·· 017
Unit 3　Bid and Construction Contracts ··································· 032
Unit 4　Project Management ·· 049
Unit 5　Geotechnical Engineering ·· 066
Unit 6　Highway Engineering ·· 081
Unit 7　Bridge Construction ·· 097
Unit 8　Rail Engineering ··· 112
Keys to Exercises ··· 126

Unit 1 Civil Engineering

Lead in

Identify the pictures below. Match them with the words in the box.

| distillation tower | crane | casting yard |

Section A

Dialogue 1

It's the first time for Mr. Gao to visit the construction site in Tanzania and Mr. Huang is going to introduce him to the other workers.

H=Mr. Huang   G=Mr. Gao   W=Mr. Wang

H: I'll take you to the construction site. Follow me, please.

G: Thank you.

H: Here we are.

G: Can you introduce me to the section chief on the building site?

H: Certainly. This is the section chief, Wang Lin, a Chinese builder.

W: How do you do? I'm glad to meet you.

G: How do you do? I'm glad to meet you, too.

W: Welcome to our building site. How long will you work here?

G: About two years. Would you introduce me to some of the workers?

W: Certainly. Hello, fellow workers! This is our new fellow, Mr. Gao.

All the workers: How do you do, Mr. Gao?

G: How do you do? Shall we start?

W: Yes, please. Here is the wood (red bricks, mixed **cement**).

G: Is that **hoist** made in Yemen?

W: Yes. But that **crane** over there is made in Japan.

G: I think, in Yemen there are many **prefabricated** houses like this building.

W: Yes. The prefabricated components are all made in the casting yard.

G: I think so.

H: Stop please. Now, time for rest.

New Words

cement	[sɪˈment] n.	a fine grey powder made of a mixture of calcined lime stone and clay 水泥
hoist	[hɔɪst] n.	any apparatus or device for hoisting 卷扬机
crane	[kreɪn] n.	a device for lifting and moving heavy object 起重机
prefabricate	[ˈpriːˈfæbrɪkeɪt] v.	to manufacture sections of (a building) 预制

Phrases and Expressions

section chief 工长

casting yard 预制厂

Dialogue 2

Mr. Du is supervising the installation of the distillation tower and Mr. Wang is in charge of the installation work.

D=Mr. Du   Y=Mr. Yang   W=Mr. Wang

D: Good Morning, everyone.

Y: Good morning, Mr. Du.

D: Today, our task is the **installation** of this distillation tower. Get ready, please.

Y: Yes. Tell us the weight, length, diameter and center of **gravity** of this tower, please.

D: The weight of this tower is thirty-four tons; length, forty-three meters; diameter, nine hundred millimeters; center of gravity, fifteen meters from the bottom of the tower.

Y: Who is the director of the installation work?

D: Engineer Wang is in charge of the installation work today. Check over the tools, such as the electrical winch, wire ropes, **pulleys**, and make sure whether all of them are in good condition. Please check the anchor bolt once more with the aid of the drawing.

W: OK. We have checked over them all.

D: Good. Begin working, please.

W: Pay attention everyone! Listen to my whistle on your post.

D: Good, the tower has been on its position, check its perpendicularity.

W: The perpendicular tolerance of the tower is less than one thousandth of its height. Acceptable!

D: OK. Tighten the anchor nuts. Have a rest, please.

New Words

installation	[ˌɪnstəˈleɪʃn] n.	the act of installing or the state of being installed 安装
gravity	[ˈɡrævəti] n.	the force that causes things to drop to the ground 重心，重力
pulley	[ˈpʊli] n.	a wheel with a grooved rim in which a rope, chain, or belt can run 滑轮

Phrases and Expressions

electrical winch 电动绞车
wire rope 钢丝绳
distillation tower 蒸馏塔
anchor bolt 基础螺丝(地脚螺栓)
anchor nuts 基础螺帽
perpendicular tolerance 垂直偏差

Exercise 1

Decide whether the following statements are true (T) or false (F) according to the dialogues.

☐ 1. The Chinese builder Wang Lin has been working at the construction site for about two years.

☐ 2. Mr. Wang is the section chief.

☐ 3. The installation tower is forty-five meters long.

☐ 4. In dialogue two, it is the engineer Wang who is responsible for the installation task.

☐ 5. The perpendicular tolerance of the installation tower is one thousandth of its height and that tolerance is totally acceptable.

Exercise 2

Oral practice.

Directions: Pair work. Use the questions below to interview your partner and then change roles.

1. Can you introduce me the section chief on the building site?

_____

_____

2. How long will you work here?

_____

_____

3. Would you introduce me some of the workers?

_____

_____

4. Who is the director of the installation work?

_____

_____

5. Is that hoist made in Yemen?

_____

_____

Exercise 3

Practical Activity.

Directions: Suppose you are Mr. Hu, an engineer from China State Construction Engineering Corporation and you are designated to take charge of the construction project in Kenya. It is the first time you meet the staff working at the construction site. Work in pairs and make a conversation.

Section B

Civil Engineering

Civil engineering, which is the oldest of the engineering specialties, refers to the planning, design, construction, and management of the built environment. This environment includes all structures built according to scientific principles, from irrigation and **drainage** systems to rocket-launching facilities.

Civil engineers build roads, bridges, tunnels, dams, harbors, power plants, water and **sewage** systems, hospitals, schools, mass transit, and other public facilities essential to modern society and large population concentrations. They also build privately owned facilities such as airports, railroads, pipelines, **skyscrapers**, and other large structures designed for industrial, commercial, or residential use. In addition, civil engineers plan, design, and build complete cities and towns, and more recently have been planning and designing space **platforms** to house **self-contained** communities.

It is traditionally broken into several sub-disciplines including environmental engineering, geotechnical engineering, structural engineering, transportation engineering, water resources engineering, community and urban planning, pipeline engineering, **photogrammetry**, surveying, mapping, and so on. In the following paragraphs, several sub-disciplines of civil engineering will be introduced **respectively**.

Structural engineering. In this specialty, civil engineers plan and design structures of all types, including bridges, dams, power plants, supports for equipment, special structures for offshore projects, the United States space program, transmission towers, giant astronomical and radio telescopes, and many other kinds of projects. Using computers, structural engineers determine the forces that a structure must resist: its own weight, wind and hurricane forces, temperature changes that expand or **contract** construction materials, and earthquakes. They also determine the combination of appropriate materials: steel, concrete, plastic, **asphalt**, brick, aluminum, and other construction materials.

Water resources engineering. Civil engineers in this specialty deal with all aspects of the physical control of water. Their projects help prevent floods, supply water for cities and for irrigation, manage and control rivers and water **runoff**, and maintain beaches and other **waterfront** facilities. In addition, they design and maintain harbors, canals, and **locks**, build huge hydroelectric dams and smaller dams and water **impoundments** of all kinds, help design offshore structures, and determine the location of structures affecting navigation.

Geotechnical engineering. Civil engineers who specialize in this field analyze the properties of soils and rocks that support structures and affect structural behavior. They evaluate and work to minimize the potential settlement of buildings and other structures that stems from the pressure of their weight on the earth. These engineers also evaluate and determine how to strengthen the stability of **slopes** and fills and how to protect structures against earthquakes and the effects of groundwater.

Environmental engineering. In this branch of engineering, civil engineers design, build and **supervise** systems to provide safe drinking water and to prevent and control pollution of water supplies, both on the surface and underground. They also design, build, and supervise projects to control or **eliminate** pollution of the land and air. These engineers build water and waste-water treatment plants, and design air **scrubbers** and other devices to minimize or eliminate air pollution caused by industrial processes, **incineration**, or other smoke-producing activities. They also work to control **toxic** and **hazardous** wastes through the construction of special dump sites or the neutralizing of toxic and hazardous substances. In addition, the engineers design and manage **sanitary** landfills to prevent pollution of surrounding land.

Transportation engineering. Civil engineers working in this specialty build facilities to ensure safety and efficient movement of both people and goods. They specialize in designing and maintaining all types of transportation facilities, highways and streets, mass transit systems, railroads and airfields, ports and harbors. Transportation engineers apply technological knowledge as well as consideration of the economic, political, and social factors in designing each project. They work closely with urban planners, since the quality of the community is directly related to the quality of the transportation system.

Pipeline engineering. In this branch of civil engineering, engineers build pipelines and related facilities which transport liquids, gases, or solids ranging from coal **slurries** (mixed coal and water) and semiliquid wastes, to water, oil, and various types of highly **combustible** and noncombustible gases. The engineers determine pipeline design, the economic and environmental impact of a project on regions it must **traverse**, the type of materials to be used – steel, concrete, plastic, or combinations of various materials – installation techniques, methods for testing pipeline strength, and controls for maintaining proper pressure and rate of flow of materials being transported. When hazardous materials are being carried, safety is a major consideration as well.

Construction engineering. Civil engineers in this field **oversee** the construction of a project from beginning to end. Sometimes called project engineers, they apply both technical and managerial skills, including knowledge of construction methods, planning, organizing, financing, and operating construction projects. They coordinate the activities of virtually everyone engaged in the work: the **surveyors**; workers who lay out and construct the temporary roads and **ramps**, **excavate** for the foundation, build the forms and pour the concrete; and workers who build the steel framework. These engineers also make regular progress reports to the owners of the structure.

Community and urban planning. Those engaged in this area of civil engineering may plan and develop community within a city, or entire cities. Such planning involves far more than engineering consideration; environmental, social, and economic factors in the use and development of land and natural resources are also key elements. These civil engineers coordinate planning of public works along with private development. They evaluate the kinds of facilities needed, including streets and highways, public transportation systems, airports, port facilities, water-supply and waste water-disposal systems, public buildings, parks, and recreational and other facilities to ensure social and economic as well as environmental **well-being**.

Photogrammetry, surveying, and mapping. The civil engineers in this specialty precisely measure the Earth's surface to obtain reliable information for locating and designing engineering projects. This practice often involves high-technology methods such as satellite and aerial surveying, and computer-processing of photographic imagery. Radio signal from satellites, scans by laser and sonic beams, are **converted** to maps to provide far more accurate measurements for **boring** tunnels, building highways and dams, **plotting** flood control and irrigation project, locating subsurface geologic formations that may affect a construction project, and a host of other building uses.

New Words

drainage	['dreɪnɪdʒ] n.	the system or process by which water or other waste liquids flow away 排水
skyscraper	['skaɪskreɪpə(r)] n.	a very tall building 摩天楼，超高层大楼
platform	['plætfɔːm] n.	a raised floor or other horizontal surface, such as a stage for speakers 平台
self-contained	[self kən'teɪnd] adj.	containing within itself all parts necessary for completeness 设备齐全的

sewage	[ˈsuːɪdʒ] n.	the waste matter carried off by sewers or drains 污水
photogrammetry	[ˌfəʊtəˈgræmətri] n.	the art or process of surveying or measuring, as in map making 摄影测量学
respectively	[rɪˈspektɪvli] adv.	in the same order as the items that you have just mentioned 分别地；各自地
contract	[ˈkɒntrækt] v.	to make or become smaller, narrower, shorter, etc. 收缩
asphalt	[ˈæsfælt] n.	substance with gravel, used in road-surfacing and roofing materials 沥青
runoff	[ˈrʌnˌɔːf] n.	sth. that runs off, as rain in excess of the amount absorbed by the ground 径流
waterfront	[ˈwɔːtəfrʌnt] n.	the area of a town or city alongside a body of water, such as a harbor 滨水地区
lock	[lɒk] n.	a section of a canal or river that may be closed off by gates to control the water level 水闸
impoundment	[ɪmˈpaʊndmənt] n.	the collection of (water) in a reservoir or dam, as for irrigation 蓄水
slope	[sləʊp] n.	an inclined portion of ground 斜坡
supervise	[ˈsuːpəvaɪz] v.	to be in charge of an activity or person 监督
eliminate	[ɪˈlɪmɪneɪt] v.	to remove or take out 消除；排除
scrubber	[ˈskrʌbə(r)] n.	an apparatus for purifying a gas 洗涤器
incineration	[ɪnˌsɪnəˈreɪʃn] n.	the act of burning something completely 焚化；烧成灰
toxic	[ˈtɒksɪk] adj.	of, relating to, or caused by a toxin or poison; poisonous 有毒的
hazardous	[ˈhæzədəs] adj.	involving great risk 有危险的；冒险的
sanitary	[ˈsænətri] adj.	of or relating to health and measures for the protection of health 卫生的
slurry	[ˈslʌri] n.	particles in a liquid, as in a mixture of cement, clay, etc. with water 泥浆
combustible	[kəmˈbʌstəbl] adj.	capable of igniting and burning 易燃的
traverse	[trəˈvɜːs] v.	to pass or go over or back and forth over (something); cross 穿过，横穿
oversee	[ˌəʊvəˈsiː] v.	to watch over and direct; supervise 监督

surveyor	[sə'veɪə(r)] n.	a person whose occupation is to survey land or buildings 测量员
ramp	[ræmp] n.	a sloping floor, path, etc. that joins two surfaces at different levels 斜坡
excavate	['ekskəveɪt] v.	to remove (soil, earth, etc.) by digging; dig out 挖掘；开凿
well-being	[ˌwel'biːɪŋ] n.	the state of being well, happy, or prosperous; welfare 幸福；康乐
convert	[kən'vɜːt] v.	to change or adapt the form, character, or function of; transform 使转变；转换
bore	[bɔː(r)] v.	to produce (a hole) by use of a drill, auger, or other cutting tool 钻孔
plot	[plɒt] v.	to draw marks or a line to represent facts, numbers etc. 绘制

Phrases and Expressions

stem from 来源于，起源于
a host of 许多，一大群

Exercise 1

Choose the best answer to each of the following questions.

1. What is **NOT** mentioned in the text?

    A. The definition of civil engineering.

    B. The scope of civil engineering.

    C. The information about the sub-disciplines of civil engineering.

    D. The ancient practices of civil engineering.

2. How many sub-disciplines of civil engineering are talked about in the text?

    A. Nine   B. Seven

    C. Eight   D. Six

3. Which specialty of civil engineering is engaged in the analysis of the properties of soils and rocks that support structures and affect structural behavior?

A. Pipeline engineering

B. Construction engineering

C. Geotechnical engineering

D. Community and urban planning

4. According to the text, which of the following statements is **NOT** true?

    A. Civil engineering has been regarded as the oldest of the engineering specialties which deals with the planning, design, construction, and management of the built environment.

    B. Those who engaged in community and urban planning help prevent floods, supply water for cities.

    C. Structural engineers make use of computers to determine the weight of a structure, wind and hurricane forces and temperature changes.

    D. Civil engineers also build privately owned facilities including airports, railroads, pipelines, skyscrapers, and other large structures.

5. Which of the following does NOT belong to the scope of environmental engineering?

    A. Building water and waste-water treatment plants.

    B. Controlling toxic and hazardous wastes through the construction of special dump sites.

    C. Evaluating water-supply and waste water-disposal systems.

    D. Designing and managing sanitary landfills.

Exercise 2

Fill in the blanks with the words given below. Change the form where necessary.

contract	supervise	eliminate	hazardous	combustible
oversee	excavate	well-being	convert	respectively

1. They are also encouraged to take on more responsibility and help _____ the newer members.

2. It is also "green", in the sense that it does not produce any _____ waste.

3. Moreover, the Treasury is an executive department, and therefore Congress and the public can more directly _____ how it uses any added authority.

4. Few things matter more to a community's _____ than the quality of its public education.

5. Of students who study in the US, the majority go to California, New York and Texas, _____ .

6. Blood is only expelled from the heart when it _____.

7. Since hydrogen gas is extremely _____, when enough hydrogen gas is mixed with air, it reacts with oxygen.

8. Recent measures have not _____ discrimination in employment.

9. By _____ the attic, they were able to have two extra bedrooms.

10. Archaeologists working on the footprints site helped him carefully _____ the tusk, which weighed 6 lbs (2.7 kg).

Exercise 3

Translate the following sentences into Chinese.

1. The network of roads on which we drive while proceeding to school or work, the huge structural bridges we come across and the tall buildings where we work, all have been designed and constructed by civil engineers.

_____

_____

2. Because it is so broad, civil engineering is subdivided into a number of technical specialties.

_____

_____

3. They may also manage private engineering firms ranging in size from a few employees to hundreds.

4. Some of the other disciplines included in civil engineering include coastal engineering, construction engineering, earthquake engineering, materials science, transportation engineering, and surveying.

5. Civil engineering has a significant role in the life of every human being, though one may not truly sense its importance in our daily routine.

课文译文

土木工程

土木工程学作为最古老的工程技术学科，是指规划、设计、施工及对建筑环境的管理。此处的环境包括建筑符合科学规范的所有结构，包括从灌溉和排水系统到火箭发射设施等。

土木工程师建造道路、桥梁、管道、大坝、海港、发电厂、给排水系统、医院、学校、公共交通和其他现代社会和大量人口集中地区的基础公共设施。他们也建造私有设施，比如飞机场、铁路、管线、摩天大楼，以及其他设计用作工业、商业和住宅途径的大型结构。此外，土木工程师还规划设计及建造完整的城市和乡镇，并且最近一直在规划设计容纳设施齐全的社区的空间平台。

传统上，我们把土木工程分为几个分支学科，包括环境工程、岩土工程、建筑工程、运输工程、社区和城市规划、水资源工程、渠道工程学和摄影测量、测量学和地图绘制等。接下来，我们将依次介绍土木工程学的分支学科。

结构工程学。在这一专业领域，土木工程师规划设计各种类型的结构，包括桥梁、大坝、发电厂、设备支撑、海面上的特殊结构、美国太空计划、发射塔、庞大的天文和无线电望远镜，以及许多其他种类的项目。结构工程师应用计算机确定一个特定结构必须承受的力：自重、风荷载和飓风荷载、建筑材料温度变化引起的胀缩，以及地震荷载。

他们也需确定不同种类的材料，如钢筋、混凝土、塑料、石头、沥青、砖、铝以及其他建筑材料等的复合作用。

水利工程学。土木工程师在这一领域主要处理水的物理控制方面的种种问题。他们的项目用于帮助预防洪水灾害，提供城市用水和灌溉用水，管理控制河流和水流物，维护河滩及其他滨水设施。此外，他们设计和维护海港、运河与水闸，建造大型水利大坝与小型坝以及各种类型的围堰，帮助设计海上结构并且确定结构的位置对航行的影响。

岩土工程学。专业于这个领域的土木工程师对支撑结构并影响结构行为的土壤和岩石的特性进行分析。他们计算建筑和其他结构由于自重压力可能引起的沉降，并采取措施使之减少到最小。他们也需计算并确定如何加强斜坡和填充物的稳定性以及如何保护结构免受地震和地下水的影响。

环境工程学。在这一工程学分支中，土木工程师设计，建造并监视系统以提供安全的饮用水，同时预防和控制地表和地下水资源供给的污染。他们也设计、建造并监视工程以控制甚至消除对土地和空气的污染。他们建造供水和废水处理厂，设计空气净化器和其他设备以最小化甚至消除由工业加工、焚化及其他产烟生产活动引起的空气污染。他们也采用建造特殊倾倒地点或使用有毒有害物中和剂的措施来控制有毒有害废弃物。此外，工程师还对垃圾掩埋进行设计和管理以预防其对周围环境造成污染。

交通工程学。从事这一专业领域的土木工程师建造可以确保人和货物安全高效运行的设施。他们专门研究各种类型运输设施的设计和维护，如公路和街道、公共交通系统、铁路和飞机场、港口和海港。交通工程师应用技术知识同时考虑经济、政治和社会因素来设计每一个项目。他们的工作和城市规划者十分相似，因为交通运输系统的质量直接关系到社区的质量。

渠道工程学。在土木工程学的这一支链中，土木工程师建造渠道和运送从煤泥浆(混合的煤和水)和半流体废污，到水、石油和多种类型的高度可燃和不可燃的气体中分离出来的液体、气体和固体的相关设备。工程师决定渠道的设计，项目所处地区必须考虑到的经济性和环境因素，以及所使用材料的类型——钢、混凝土、塑料或多种材料的复合的安装技术，测试渠道强度的方法，和控制所运送流体材料保持适当的压力和流速。当流体中携带危险材料时，安全性因素也需要考虑。

建筑工程学。土木工程师在这个领域中从开始到结束监督项目的建筑。他们，有时被称为项目工程师，应用技术和管理技能，包括建筑工艺、规划、组织、财务和操作项目建设的知识。事实上，他们协调工程中每个人的活动：测量员，布置和建造临时道路和斜坡、开挖基础、支模板和浇注混凝土的工人，以及钢筋工人。这些工程师也向结构的业主提供进度计划报告。

社区和城市规划。从事土木工程这一方面工作的工程师可能规划和发展一个城市中的社区，或整个城市。此规划中所包括的远远不止工程因素，土地的开发使用和自然资源中的环境、社会和经济因素也是主要的成分。这些土木工程师对公共建设工程的规划和私人建筑的发展进行协调。他们评估所需的设施，包括街道、公路、公共运输系统、机场、港口、给排水和污水处理系统、公共建筑、公园和娱乐及其他设施以保证社会、经济和环境的协调发展。

摄影测量，测量学和地图绘制。在这一专业领域的土木工程师精确测量地球表面以获得可靠的信息来定位和设计工程项目。这一方面包括高工艺学方法，如卫星成像、航拍和计算机成像。来自人造卫星的无线电信号，通过激光和音波柱扫描被转换为地图，为隧道钻孔，建造高速公路和大坝，绘制洪水控制和灌溉方案，定位可能影响建筑项目的地下岩石构成，以及许多其他建筑用途提供更精准的测量。

References

[1] 李锦辉,陈锐. 土木工程专业英语[M]. 上海：同济大学出版社，2012.

Unit 2　Civil Engineering Materials

Lead in

Identify the pictures below. Match them with the words in the box.

| concrete | cement | aggregate |

Section A

Dialogue 1

Steven, an intern journalist, is going to interview some working men. Here is the conversation between Steven and one of his interviewees Wang Lei.

S=Steven　　W=Wang Lei

S: Excuse me, could you please spare some time for me? I am an intern journalist. I have been assigned a task to interview some working men.

W: Yes, of course.

S: So may I ask what your profession is?

W: My profession is about civil engineering. And I am a **builder**.

S: How many years have you been working?

W: For ten years.

S: Would you please tell me more details about your work?

W: Okay.

S: Er, what is your work mainly about?

W: Our work mainly involves **construction**, such as the construction of **skyscrapers**, roads, **foundation**s and reinforced concrete structures, etc.

S: Thank you very much.

New Words

builder	[ˈbɪldə(r)] n.	a person whose job is to build or repair houses 建筑工人
construction	[kənˈstrʌkʃn] n.	the process of building or making roads, etc. 建筑
skyscraper	[ˈskaɪskreɪpə(r)] n.	a very tall building in a city 摩天大楼
foundation	[faʊnˈdeɪʃn] n.	a layer of bricks, concrete, etc. that forms the solid underground base of a building 基础

Phrases and Expressions

intern journalist 实习记者
civil engineering 土木工程
reinforced concrete structure 钢筋混凝土结构

Dialogue 2

Tom, a professor of civil engineering department, is going to teach something about civil engineering materials. Here is the conversation between him and one of the students Mike.

T=Tom   M=Mike

T: Today, we are going to learn something about civil engineering materials. First of all, can you tell me some important construction materials used in civil engineering?

M: Cement, **concrete**, steel, **aggregate** and **timber**, etc.

T: Well done. As you mentioned cement and concrete, do you know how they are made?

M: No, I don't know clearly.

T: Don't worry then. In today's class, we'll learn much more details about the chemical process of manufacturing cement. After today's class, you will be sent to the cement and concrete plants to know more about how to make cement and concrete.

M: Okay, I am looking forward to learning from the **site**.

New Words

concrete	[ˈkɒnkriːt] n.	building material made by mixing cement, sand, small stones and water 混凝土
aggregate	[ˈægrɪgət] n.	sand or broken stone that is used to make concrete or for building roads, etc. 骨料
timber	[ˈtɪmbə(r)] n.	wood that is prepared for use in building, etc. 木材
site	[saɪt] n.	a place where a building, town, etc. was, is or will be located 现场

Phrases and Expressions

cement and concrete plant 水泥和混凝土工厂

Exercise 1

Decide whether the following statements are true (T) or false (F) according to the dialogue.

☐ 1. The intern journalist is assigned to interview some construction workers.

☐ 2. Steven has been working for ten years.

☐ 3. Wang Lei's work mainly involves construction.

☐ 4. Mike already knows something about making cement.

☐ 5. Cement, concrete and aggregate are important construction materials.

Exercise 2

Oral practice.

Directions: Pair work. Use the questions below to interview your partner and then change roles.

1. How much do you know about the civil engineering materials?

_____

_____

2. What do construction workers usually do in their daily job?

_____

_____

3. Do you know the process of making cement?

_____

_____

4. Have you ever been to the construction site?

_____

_____

5. Compared with the knowledge learnt from the textbook, do you think knowledge learnt from the construction site is more important?

_____

_____

Exercise 3

Practical Activity.

Directions: Work in pairs. Suppose you are assigned to buy some important civil engineering materials from a foreign company, how are you going to make the purchase? The result of the discussion will be reported to the whole class.

Section B

How Cement and Concrete are Made

Cement is manufactured through a closely controlled chemical combination of **calcium**, **silicon**, **aluminum**, iron and other ingredients.

Common materials used to manufacture cement include **limestone**, **shells**, and **chalk** or **marl** combined with **shale**, **clay**, **slate**, blast furnace slag, silica sand, and iron ore. These ingredients, when heated at high temperatures form a rock-like substance that is ground into the fine powder that we commonly think of as cement.

The most common way to manufacture Portland cement is through a dry method. The first step is to **quarry** the principal raw materials, mainly limestone, clay, and other materials. After quarrying the rock is **crushed**. This involves several stages. The first crushing reduces the rock to a maximum size of about 6 inches. The rock then goes to secondary crushers or hammer mills for reduction to about 3 inches or smaller.

The crushed rock is combined with other ingredients such as iron ore or fly ash and **ground**, mixed, and fed to a cement **kiln**.

The cement kiln heats all the ingredients to about 2,700 degrees Fahrenheit in huge **cylindrical** steel **rotary** kilns lined with special **firebrick**. Kilns are frequently as much as 12 feet in diameter—large enough to accommodate an automobile and longer in many instances than the height of a 40-story building.

The finely ground raw material or the **slurry** is fed into the higher end. At the lower end is a **roaring** blast of flame, produced by precisely controlled burning of powdered coal, oil, alternative fuels, or gas under forced draft.

As the material moves through the kiln, certain elements are driven off in the form of gases. The remaining elements **unite** to form a new substance called **clinker**. Clinker comes out of the kiln as grey balls, about the size of **marbles**.

Clinker is **discharged** red-hot from the lower end of the kiln and generally is brought down to handling temperature in various types of **coolers**. The heated air from the coolers is returned to the kilns, a process that saves fuel and increases burning efficiency.

After the clinker is cooled, cement plants grind it and mix it with small amounts of **gypsum** and limestone. Cement is so fine that one pound of cement contains 150 billion grains. The cement is now ready for transport to ready-mix concrete companies to be used in a variety of construction projects.

Concrete is a mixture of **paste** and **aggregates**, or rocks. The paste, composed of Portland cement and water, coats the surface of the fine (small) and coarse (larger) aggregates. Through a chemical reaction called **hydration**, the paste hardens and gains strength to form the rock-like mass known as concrete.

Within this process lies the key to a remarkable trait of concrete: it's plastic and **malleable** when newly mixed, strong and **durable** when hardened. These qualities explain why one material, concrete, can build skyscrapers, bridges, sidewalks and superhighways, houses and dams.

The key to achieve a strong, durable concrete rests in the careful proportioning and mixing of the ingredients. A mixture that does not have enough paste to fill all the **voids** between the aggregates will be difficult to place and will produce rough surfaces and **porous** concrete. A mixture with an excess of cement paste will be easy to place and will produce a smooth surface; however, the resulting concrete is not cost-effective and can more easily crack.

Portland cement's chemistry comes to life in the presence of water. Cement and water form a paste that coats each particle of stone and sand—the aggregates. Through a chemical reaction called hydration, the cement paste hardens and gains strength.

The quality of the paste determines the character of the concrete. The strength of the paste, in turn, depends on the **ratio** of water to cement. The water-cement ratio is the weight of the mixing water divided by the weight of the cement. High-quality concrete is produced by lowering the water-cement ratio as much as possible without sacrificing the workability of fresh concrete, allowing it to be properly placed, **consolidated**, and **cured**.

A properly designed mixture possesses the desired workability for the fresh concrete and the required durability and strength for the hardened concrete. Typically, a mix is about 10 to 15 percent cement, 60 to 75 percent aggregate and 15 to 20 percent water. Entrained air in many concrete mixes may also take up another 5 to 8 percent.

Almost any natural water that is drinkable and has no pronounced taste or odor may be used as mixing water for concrete. Excessive **impurities** in mixing water not only may affect setting time and concrete strength, but can also cause **efflorescence**, **staining**, corrosion of reinforcement, volume instability, and reduced durability. Concrete mixture specifications usually set limits on **chlorides, sulfates, alkalis,** and solids in mixing water unless tests can be performed to determine the effect the impurity has on the final concrete.

Although most drinking water is suitable for mixing concrete, aggregates are chosen carefully. Aggregates comprise 60 to 75 percent of the total volume of concrete. The type and size of aggregate used depends on the thickness and purpose of the final concrete product.

Relatively thin building sections call for small coarse aggregate, though aggregates up to six inches in diameter have been used in large dams. A continuous **gradation** of particle sizes is desirable for efficient use of the paste. In addition, aggregates should be clean and free from any matter that might affect the quality of the concrete.

Soon after the aggregates, water, and the cement are combined, the mixture starts to harden. All Portland cements are hydraulic cements that set and harden through a chemical reaction with water call hydration. During this reaction, a **node** forms on the surface of each cement particle. The node grows and expands until it links up with nodes from other cement particles or adheres to **adjacent** aggregates.

Once the concrete is thoroughly mixed and workable it should be placed in forms before the mixture becomes too stiff.

During placement, the concrete is consolidated to **compact** it within the forms and to eliminate potential flaws, such as **honeycombs** and air pockets.

New Words

calcium	['kælsɪəm] n.	a soft silver-white metal found in bones, teeth and chalk 钙
silicon	['sɪlɪkən] n.	a chemical element existed as a gray solid or brown powder and found in rocks and sand 硅
aluminum	[ə'lju:mɪnəm] n.	a light, silver-gray metal used for making pans, etc. 铝
limestone	['laɪmstəʊn] n.	a type of white stone that contains calcium, used in building and in making cement 石灰岩
shell	[ʃel] n.	a gray powder made by burning clay and lime that sets hard when it is mixed with water 壳
chalk	[tʃɔ:k] n.	a type of soft white stone 白垩
marl	[mɑ:l] n.	soil consisting of clay and lime 泥灰

shale	[ʃeɪl] n.	a type of soft stone that splits easily into thin flat layers 页岩
clay	[kleɪ] n.	a type of heavy, sticky earth that becomes hard when baked and used to make pots and bricks 粘土
slate	[sleɪt] n.	a type of dark gray stone that splits easily into thin flat layers 板岩
quarry	[ˈkwɒri] n. v.	a place where large amounts of stone, etc. are dug out of the ground 采石场 to take stone, etc. out of a quarry 开采
crush	[krʌʃ] v.	to break sth. into small pieces or into a powder by pressing hard 压碎
grind	[ɡraɪnd] v.	to break or crush sth. into very small pieces 磨细
kiln	[kɪln] n.	a large oven for baking clay and bricks, drying wood and grain, etc. （烧制砖头等的）窑
cylindrical	[səˈlɪndrɪkl] adj.	shaped like a cylinder 圆柱形的
rotary	[ˈrəʊtəri] adj.	moving in a circle around a central fixed point 旋转的
firebrick	[ˈfaɪəbrɪk] n.	brick which is not destroyed by very strong heat 耐火砖
slurry	[ˈslʌri] n.	a thick liquid consisting of water mixed with animal waste, clay, coal dust or cement 泥浆
roaring	[ˈrɔːrɪŋ] adj.	burning with a lot of flames and heat 熊熊燃烧的
unite	[juˈnaɪt] v.	to join together 联合
clinker	[ˈklɪŋkə(r)] n.	the hard rough substance left after coal has burnt at a high temperature 煤渣
marble	[ˈmɑːbl] n.	a small ball of colored glass that children roll along the ground in a game 玻璃弹珠
discharge	[dɪsˈtʃɑːdʒ] v.	flows somewhere 排出
cooler	[ˈkuːlə(r)] n.	a container or machine which cools things, especially drinks, or keeps them cold 冷却器
gypsum	[ˈdʒɪpsəm] n.	a soft white mineral like chalk found naturally and used in making plaster of Paris 石膏
paste	[peɪst] n.	a soft wet mixture, usually made of a powder and a liquid 泥浆
hydration	[haɪˈdreɪʃn] n.	the process of absorbing water 水化

malleable	['mælɪəbl] adj.	that can be hit or pressed into different shapes easily without breaking or cracking 可锻的
durable	['djʊərəbl] adj.	likely to last for a long time without breading 耐久的
void	[vɔɪd] n.	a large empty space 空隙
porous	['pɔːrəs] adj.	having many small holes that allow water or air to pass through slowly 多孔的
ratio	['reɪʃɪəʊ] n.	relationship between two groups of people or things represented by two numbers showing how much larger one is than the other 比例
consolidate	[kən'sɒlɪdeɪt] v.	to make stronger 固结
cure	[kjʊə(r)] v.	to treat with smoke, salt, etc. in order to preserve it 养护
impurity	[ɪm'pjʊərəti] n.	a substance that is present in small amounts in another substance, making it dirty or of poor quality 杂质
efflorescence	[ˌeflə'resns] n.	the powder which appears on the surface of bricks, etc. when water evaporates 渗斑
staining	[steɪnɪŋ] n.	to leave a mark that is difficult to remove 滤除
chloride	['klɔːraɪd] n.	compound of chlorine and another chemical element 氯化物
sulfate	['sʌlˌfeɪt] n.	a salt that is formed when sulfuric acid reacts with another chemical element 硫酸钠
alkali	['ælkəlaɪ] n.	a chemical substance that reacts with acids to form a salt 碱
gradation	[grə'deɪʃn] n.	the process of sth. changing gradually 级配
node	[nəʊd] n.	a point two lines or systems meet or cross 节点
adjacent	[ə'dʒeɪsnt] adj.	next to or near 临近的
compact	[kəm'pækt] v.	to press sth. together firmly 压实
honeycomb	['hʌnɪkəʊm] n.	a structure of cells with six sides, made by bees for holding their honey and their eggs 蜂巢状

Phrases and Expressions

blast furnace slag 高炉渣

iron ore 铁矿石

be ground into 被磨碎成

fine powder 细粉

Portland cement 波特兰水泥，普通水泥

secondary crusher 复轧碎石机

hammer mill 锤片式粉碎机

fly ash 粉煤灰

forced draft 强制通风

entrain air 携入的空气

pronounced taste 明显的味道

corrosion of reinforcement 钢筋腐蚀

volume instability 体积不稳

hydraulic cement 水硬水泥

air pocket 鼓泡

Exercise 1

Choose the best answer to each of the following questions.

1. Which of the following is **NOT** one of the common materials used to make cement?

    A. Clay   B. Limestone

    C. Chalk   D. Silicon

2. Which of the following is true about clinker?

    A. Clinker is driven off in the form of gases.

    B. Clinker is like the marbles.

    C. Clinker flows out from the lower end of the kiln.

    D. Clinker is used to handle various coolers.

3. Why concrete can be used to build skyscrapers, bridges, sidewalks and superhighways, houses and dams?

    A. Because it is newly mixed.

    B. Because it is hardened.

C. Because of all its qualities.

D. Because it is an important building materials.

4. The following materials are all used in manufacturing concrete **EXCEPT** _____.

   A. Mixing water

   B. Cement

   C. Aggregate

   D. Entrained air

5. What factors decide the type and size of the used aggregate?

   A. The thinness of the final concrete product.

   B. The aim and thickness of the final concrete product.

   C. The purpose of the final concrete product.

   D. The thinness and purpose of the final concrete product.

Exercise 2

Fill in the blanks with the words given below. Change the form where necessary.

| unite | discharge | grind | durable | porous |
| ratio | consolidate | compact | adjacent | fin |

1. We will _____ in fighting crime.

2. The administration hopes that such measures will _____ its position.

3. The cement need not be finely _____.

4. The _____ of applications to available places currently stands at 100:1.

5. He added sand to the soil to make it more _____.

6. The snow had _____ into a hard icy layer.

7. The river is diverted through the power station before _____ into the sea.

8. There is a row of houses immediately _____ to the factory.

9. Acupuncture uses _____ needles inserted into the patient's skin.

10. Painted steel is likely to be less _____ than other kinds.

Exercise 3

Translate the following sentences into Chinese.

1. Kilns are frequently as much as 12 feet in diameter—large enough to accommodate an automobile and longer in many instances than the height of a 40-story building.

2. The heated air from the coolers is returned to the kilns, a process that saves fuel and increases burning efficiency.

3. Through a chemical reaction called hydration, the paste hardens and gains strength to form the rock-like mass known as concrete.

4. A mixture with an excess of cement paste will be easy to place and will produce a smooth surface; however, the resulting concrete is not cost-effective and can more easily crack.

5. Relatively thin building sections call for small coarse aggregate, though aggregates up to six inches in diameter have been used in large dams.

课文译文

水泥和混凝土是如何制成的

水泥是通过严密控制的钙、硅、铝、铁和其他原料的化学结合而制成的。

用于制作水泥的常见材料包括石灰岩、壳和白垩或者泥灰,并混合页岩、黏土、板岩、高炉渣、硅砂和铁矿石。经过高温加热后,这些成分会形成一种形似岩石的物质。这种物质被磨碎成细粉状,而这种细粉状物质我们通常把它叫作水泥。

普通水泥最常见的制作方式是干法制作。第一步是开采主要的原材料,主要是石灰岩、黏土和其他的材料。开采之后,岩石被压碎。这一步涉及几个阶段。首次压碎将岩石的最大尺寸减小到大约 6 英寸。接着,岩石进入复轧碎石机或者锤片式粉碎机,尺寸减小到大约 3 英寸或者更小。

被压碎的岩石,添加其他诸如铁矿石或者粉煤灰等成分,被磨细、混合,并被送进水泥窑里。

水泥窑在添加特殊耐火砖作为内衬的大型圆柱钢回转窑里将所有的成分加热到大约 2 700 华氏度。水泥窑通常直径宽达 12 英尺——大到足以容纳一辆汽车。并且,在许多情况下,水泥窑比一栋 40 层高的建筑物的高度还要长。

被磨细的原材料或者泥浆被送进较高的地方,下端是熊熊燃烧的火焰。火焰是在强制通风的情况下由于精密控制而燃烧的粉末状煤块、燃油、替代性燃料或者燃气而产生的。

当材料穿过水泥窑时,特定的成分以气体的形式被驱散出去。剩下的成分联合形成一种叫作煤渣的新物质。煤渣以灰色球体的形式从水泥窑里散出来,大约是玻璃弹珠的大小。

滚烫的煤渣从水泥窑的下端被排出,并在各种类型的冷却器中进行温度处理。冷却器里的热空气回到水泥窑里,这个过程能够节约燃料并增加燃烧率。

在煤渣冷却后,水泥厂将其磨细,并添加少量的石膏和石灰岩加以混合搅拌。水泥非常细,一磅水泥有 1 500 亿颗粒。现在,水泥已经准备好被运往预拌混凝土公司,可以将其用于各种建筑工程中。

混凝土是水泥浆和骨料,或者岩石的混合搅拌。水泥浆是由普通水泥和水混合而成的,并覆盖在细、粗骨料表面上。通过一个被称作水化的化学反应,水泥浆变硬,硬度增强,形成一种像岩石的物质,这种物质被称为混凝土。

这个过程凸显了混凝土一个显著的特征的关键:当才被混合搅拌时,混凝土是可塑和可锻的,而当变硬时,混凝土是坚固持久的。这些特性正好说明了为什么混凝土这一种材料可以建造摩天大楼、桥梁、人行道、高速公路、房屋和水坝。

实现制作坚固耐久的混凝土的关键在于原料精细的比例和混合。如果混合搅拌物没有足够多的水泥浆来填充骨料之间所有的空隙，将会难以被浇筑，同时，会造成混凝土表面粗糙形成多孔混凝土。如果混合搅拌物的水泥浆过量，将会很容易浇筑，并制造出一个光滑的混凝土表面；但是，所生产出来的混凝土成本不低，而且很容易破裂。

普通水泥的化学过程遇水便变得活跃起来。水泥和水形成水泥浆，覆盖在每一粒石头和沙子——骨料上。通过一种叫作水化的化学反应，水泥浆硬化，硬度增强。

水泥浆的质量决定了混凝土的特性。反过来，水泥浆的强度取决于水和水泥的比例。水灰比是拌合水的重量除以水泥的重量。优质混凝土是通过尽可能多地降低水灰比来制成的，而没有牺牲掉新拌混凝土的和易性，使得混凝土能够被恰当地浇筑、固结和养护。

恰当设计好的搅拌物有着新拌混凝土理想的和易性，以及硬化混凝土所需要的耐久性和硬度。一般情况下，搅拌物里大约有10%~15%的水泥、60%~75%的骨料和15%~20%的水。在很多混凝土搅拌物中被携入的空气或许还会占据5%~8%的比例。

几乎任何一种可以喝、没有明显味道和气味的天然水都可以被用作混凝土的拌合水。拌合水里过度的杂质不仅会影响凝结时间和混凝土强度，而且还会造成渗斑、滤除、钢筋腐蚀、体积不稳以及耐久性降低。混凝土搅拌物规格通常会限制拌合水里的氯化物、硫酸钠、碱和固体物，除非通过测试来决定杂质对于最后制成的混凝土的影响。

虽然大部分饮用水都可以被用于搅拌混凝土，但是骨料须得精心挑选。骨料占据了混凝土总量的60%~75%。使用的骨料的类型和尺寸取决于最后制成的混凝土的厚度和用处。

相对较薄的建筑部分需要小且粗的骨料，而直径宽达6英寸的骨料常被用于建造大型水坝。持续性的颗粒尺寸级配是有效利用水泥浆所需要的。而且，骨料应该干净，没有添加任何其他物质，因为这些物质或许会影响到混凝土的质量。

一旦骨料、水和水泥被混合搅拌后，搅拌物开始硬化。所有的普通水泥都是水硬水泥通过一个和水的叫作水化的化学反应固定并硬化而成的。在这个化学反应过程中，在每一个水泥粒表面上会形成一个节点。这个节点变大并膨胀，直到它和其他水泥粒的节点连在一起或者黏附在临近的骨料上。

一旦混凝土被彻底搅拌并可塑，它就应该在变得太过坚硬前被浇筑到模壳里。

在浇筑的过程中，混凝土被加固，在模壳里被压实，并消除可能的瑕疵，如蜂窝状和鼓泡。

References

[1] America's Cement Manufacturers. The Portland Cement Association[EB/OL]. (2017-02-16) http://www.cement.org/cement-concrete-basics/how-concrete-is-made.

[2] Civil Engineering Materials [EB/OL]. (2017-02-05) http://www.doc88.com/p-683738848053.html

Unit 3 Bid and Construction Contracts

Lead in

Identify the pictures below. Match them with the words in the box.

| contract | cash deposit | bidder |

[   ]   [   ]   [   ]

Section A

Dialogue 1

Li Yun, a representative of China Star Construction Company limited, is going to inquire some information about a bid.

L=Li Yun   J=Johnson

L: Good morning, Mr. Johnson. I'm Li Yun, from China Star Construction Company

Limited. I've heard that your company is prepared to call for a bid for a commercial building construction project in Singapore, is that true?

J: Yes, we are making arrangements these days. You are well informed.

L: Is this **tender**-opening done publicly?

J: Yes, tender-opening is done publicly this time, all the bidders shall be invited to join us to **supervise** the opening.

L: What's the time set for the bidders to **submit** their bids?

J: The time period for bidding is temporarily set from May 1st to the end of May. We'll **release** the notice and construction documents on the official website of our company and give the **bidders** 4 weeks to prepare their bids.

L: Should the bidders make a cash **deposit**?

J: Yes, certainly. According to the international practice, bidders need to hand in a cash deposit or a letter of guarantee from a commercial bank. If one fails to win the **award**, the cash deposit or the letter of guarantee shall be returned to the bidder within one week after the decision on award is **declared**.

L: Our corporation is very interested in this tender. We will try our best to win the award.

J: I understand fully how you feel. If the conditions of your tender prove to be most suitable, of course we'll accept your tender.

New Words

bid	[bɪd] n.	a formal proposal to buy at a specified price 出价，投标
tender	['tendə(r)] n.	a formal offer to carry out work at a stated price 招标；投标
supervise	['su:pəvaɪz] vt.& vi.	to keep an eye on 监督；管理
submit	[səb'mɪt] vt.	to give a document, proposal, etc. in authority 提交，呈送
release	[rɪ'li:s] vt.	to distribute an announcement to the public 发布；发行
bidder	['bɪdə(r)] n.	a person or group that offers to do sth in competition with others 投标人

deposit	[dɪˈpɒzɪt] n.	money given as security for an article acquired for temporary use 保证金
award	[əˈwɔːd] n.	the official decision to give sth. to sb. 授予；裁定
declare	[dɪˈkleə(r)] vt.	to say sth officially or publicly 宣布；公布

Phrases and Expressions

call for a bid 招标

tender-opening 开标

make a cash deposit 缴纳保证金

Dialogue 2

Li Yun, a representative of China Star Construction Company Limited, is going to sign the contract with Johnson.

L=Li Yun   J=Johnson

L: Good morning, Johnson.

J: Good morning, Mr. Li.

L: We've brought the **draft** of our contract. Please have a look.

J: We have reached a basic agreement on the prices and the problems that should be worked out.

L: Both of our parties have made a great effort.

J: Well, the most important thing is that our company demands the quality be exactly the same as the terms and conditions **stipulated** in this contract.

L: Please feel **assured** that we'll abide by our promise.

J: One more thing, the time limit for the project is very urgent. The whole project should be completed before March 1, 2019. Do you have any difficulties?

L: We'll do everything we can to ensure **delivery**.

J: That's fantastic! It's time for us to sign the contract.

L: There're the **originals** of the contract, please check it carefully.

J: Fine. I'd like to go over it. (After about 15 minutes) Hmm, no problem.

L: I've been looking forward to this moment. Now, please **countersign** it.

J: Done. Congratulations.

L: Thank you very much. I think the contract will bear fruit in no time, and I expect our continuing cooperation.

New Words

contract	['kɒntrækt] n.	a binding agreement between two or more persons that is enforceable by law 合同；协议
draft	[drɑ:ft] n.	a rough written version of sth that is not yet in its final form 草案
stipulate	['stɪpjuleɪt] vt.	(formal) to state clearly and firmly that sth must be done（尤指在协议或建议中）规定，约定
assured	[ə'ʃʊəd] adj.	certain to happen 确定的
delivery	[dɪ'lɪvəri] n.	the act of taking goods, letters, etc. to the people they have been sent to（正式）交付
original	[ə'rɪdʒənl] n.	a document produced for the first time, from which copies are later made 正本
countersign	['kaʊntəsaɪn] vt.	to sign a document that has already been signed by another person [术]连署，副署，会签（文件）

Phrases and Expressions

reach an agreement 达成协议

terms and conditions 条件条款

abide by 遵守；信守

sign the contract 签署合同

bear fruit 结出果实，奏效

in no time 马上，立即

Exercise 1

Decide whether the following statements are true (T) or false (F) according to the dialogues.

☐ 1. The bid in Dialogue 1 is invited bidding instead of open bidding.

☐ 2. The bidders should submit their bids before May 1st.

☐ 3. The cash deposit can be refunded to the bidder if he does not win the tender.

☐ 4. In Dialogue 2, the two parties made some amendments to the contract.

☐ 5. Johnson attached great importance to the quality of the project.

Exercise 2

Oral practice.

Directions: Pair work. Use the questions below to interview your partner and then change roles.

1. When are you scheduled to open the tender and where?

_____

_____

2. Should the bidders make a cash deposit?

_____

_____

3. Would you please let me know something more about your conditions for this tender?

_____

_____

4. Are there any special requirements on the project?

_____

_____

5. Can you guarantee delivery in time?

_____

_____

Exercise 3

Practical Activity.

Directions: Work in pairs. Suppose you are an employee of China Star Construction Company limited, and your company has won the tender to build a railway project in South Africa. As a representative of your company, you are going to TM International Investment Corporation to negotiate some details and finally sign the contract. The result of the discussion will be reported to the whole class.

Section B

Contract of Construction

Party A: TM International Investment Corporation

Party B: China Star Construction Company Limited

This contract is hereby signed by the two Parties concerned in Singapore on October 15, 2016 according to the *Contract Law of the People's Republic of China*, *Construction Law of the People's Republic of China*, the *Regulation on Building and* **Installation** *Contracting Contract* and other relevant laws and regulations on the principles of equality, voluntariness, fairness and good faith.

1. Project Introduction:

1) Name of Project: Modern Commercial Building

2) Location of Project: in Singapore

3) Project content: Twenty-five floors with **reinforced** concrete structure; Construction area: 24, 240 square meters

4) Project approval No.: 1617010023

5) Capital source: Foreign capital

6) Date of the project beginning: March 1, 2017

　　Date of the project completion: March 1, 2019

7) Contract Amount: The total contract amount is 100,000,000 Yuan, which is fixed contract amount.

2. Contracting Scope

1) Civil engineering

2) Inner decoration

3) **Facade** (including doors, windows, roof and stairs)

4) Electronic engineering

5) Air conditioning system

6) Water supply, **drainage** and heating system

7) Fire control system

3. Contract Documents

Documents forming part of the Contract include:

1) The Contract Agreement

2) Bid-winning Notice

3) Tender Document and Its Appendix

4) Standard, Specifications and Relevant Technical Documents

5) Drawings

6) Bill of Quantities

7) Engineering Bid or **Budget** Statement

4. Rights and Obligations of Party A

(1) Party A shall provide five copies of confirmed construction drawings and explanations of work procedures, and shall **conduct** technical **clarification**. Party A shall provide Party B with necessary water and power equipment and the instructions on using the equipment.

(2) Party A appoints the on-site representative to supervise the **fulfillment** of contract obligations, to **oversee** the quality control and check the project progress, and to handle issues such as acceptance and variations. In the event of the on-site representative of Party A believes that any employee of Party B has improper conduct or misbehaviors, or if Party A consider such employee is not competent for the relevant work, Party B shall **dismiss** and replace this employee immediately.

(3) In the event of Party A consider Party B is indeed unable to further fulfill its

contract obligations, then Party A has the right to **terminate** the contract, and Party B should **vacate** the project site within two weeks after receiving Party A's notice, so that Party A can **resume** construction as soon as possible. Confirmation of finished quantity shall not be regarded as a **prerequisite** to Party B's vacation of the project site, and the confirmation should be processed within one month after the project completion.

(4) Party A should make payments to party B according to the payment terms and conditions agreed in this Contract. Payment terms: Party A shall pay 30% of the total contract amount to Party B within 5 working days after signing the contract; 20% of the total contract amount shall be paid within 5 working days after the facade is finished; 20% of the total contract amount shall be paid within 5 working days after the construction mid-term check and acceptance is completed; 30% of the total contract amount shall be paid within 5 workings after the construction completion check and acceptance.

5. Rights and Obligations of Party B

(1) Party B shall participate in the technical clarification on construction drawings and explanations of work procedures organized by Party A, and should develop the construction plan and the progress schedule for Party A to review. During construction, Party B should make such adjustment to the construction plan and the progress schedule per Party A's instruction and change-orders.

(2) Party B appoints the on-site representative of Party B to take charge of fulfilling the contract obligations. The on-site representative of Party B shall organize the construction according to contract requirements so that the project can be completed according to the quality and quantity requirements and on a timely manner. The on-site representative of Party B shall work on the project site for no fewer than 90% of the total working days.

(3) Party B shall carry out the construction according to the design drawings agreed upon by both Party A and the designer, shall **observe** the instructions of designer and the representative of Party A; in the event of Party B finds **inconsistence** between project standards and design drawings, Party B should inform the designer and Party A in writing.

(4) From the beginning of construction to the completion date stipulated by the as-built acceptance report, Party B will be fully responsible for the entire project. Party B shall fix and **reinstate** the damages to the project items regardless of the cause, so that the project shall be in good condition upon completion, according to the contract requirements and the

instructions of Party A is representative.

(5) Party B should sign labor contract with all its construction staffs, issue legal labor certificate and be responsible for the corresponding cost. Party B should make salary/wages payment to all construction staffs in time on a monthly basis, and will be fully responsible for any **liabilities** caused by any loss or damage to Party A as a result of Party B's failure to pay salaries and wages according to the requirements.

6. Force Majeure

Party B is not liable for failure to perform the contractor's obligations if such failure is as a result of Force Majeure such as fire, flood, government forces and other Force Majeure factors. However, Party B shall timely notice Party A of the occurrence of Force Majeure and provide the supporting documents of relevant government authorities.

7. Dispute Settlement

All disputes arising in connection with this contract or the execution thereof shall be settled by way of friendly negotiation. In case no settlement can be reached, the case at issue shall then be submitted for **arbitration** to the China International Economic and Trade Arbitration commission in accordance with the **provision**s of the said commission. The award by the said commission shall be **deem**ed as final and **binding** upon both parties.

8. Execution of Contract

This contract is written in both Chinese and English in quadruplicate, with both texts being equally authentic and each Party shall hold two copies. The contract becomes effective from the date of signing and stamping by both Parties. In the event of any conflict between the English and Chinese, the latter shall **prevail**.

Party A: Party B:

Date: Date:

New Words

installation	[ˌɪnstəˈleɪʃn] n.	the act of fixing equipment or furniture in position so that it can be used 安装
reinforce	[ˌriːɪnˈfɔːs] vt.	to make a structure or material stronger 加固
facade	[fəˈsɑːd] n.	the face or front of a building （建筑物的）正面，立面
drainage	[ˈdreɪnɪdʒ] n.	the process by which water or liquid waste is drained from an area 排水；放水
budget	[ˈbʌdʒɪt] n.	a sum of money allocated for a particular purpose 预算
obligation	[ˌɒblɪˈgeɪʃn] n.	the state of being forced to do sth because it is your duty 义务，责任
conduct	[kənˈdʌkt] vt.	to organize and/or do a particular activity 实施；执行
clarification	[ˌklærəfɪˈkeɪʃn] n.	(formal) to make sth clearer to understand 交底
fulfillment	[fʊlˈfɪlmənt] n.	to do what is required or necessary 履行；执行
oversee	[ˌəʊvəˈsiː] vt.	to watch sb/sth and make sure that a job or an activity is done correctly 监督，监视
dismiss	[dɪsˈmɪs] vt.	to officially remove sb from their job 解雇，把…免职
terminate	[ˈtɜːmɪneɪt] vt.&vi.	to end; to make sth end 结束；使终结
vacate	[vəˈkeɪt] vt.	to leave a building, seat, etc., especially so that sb else can use it 搬出；使撤退
resume	[rɪˈzjuːm] v.	take up or begin anew 重新开始；继续
prerequisite	[ˌpriːˈrekwəzɪt] n.	sth that is required in advance 先决条件，前提
observe	[əbˈzɜːv] vt.	to obey rules, laws, etc. 遵守
inconsistence	[ɪnkənˈsɪstəns] n.	not matching a set of standards, ideas 不一致
reinstate	[ˌriːɪnˈsteɪt] vt.	restore to the previous state 使复原；使恢复
liability	[ˌlaɪəˈbɪlɪti] n.	the state of being legally responsible for sth 责任，债务
arbitration	[ˌɑːbɪˈtreɪʃn] n.	the official process of settling an argument or a disagreement 仲裁，公断
provision	[prəˈvɪʒn] n.	a condition or an arrangement in a legal document 规定，条款

commission	[kəˈmɪʃn] n.	an official group of people who have been given responsibility to control sth 委员会
deem	[di:m] vt.	to have a particular opinion about sth 视为
binding	[ˈbaɪndɪŋ] adj.	sth that must be obeyed because it is accepted in law 有约束力的；应履行的
execution	[ˌeksɪˈkju:ʃn] n.	the act of doing a piece of work, performing a duty, or putting a plan into action 执行
authentic	[ɔːˈθentɪk] adj.	sth that is legally or officially acceptable [法]认证了的
prevail	[prɪˈveɪl] vi.	to be accepted, especially after a struggle or an argument 获胜，占优势

Phrases and Expressions

in the event of 倘若，如果
be competent for 胜任
make adjustment to 作出调整
take charge of 负责；监理
carry out 执行；进行

Exercise 1

Choose the best answer to each of the following questions.

1. In this contract, Party B is _____?
   A. The Employer
   B. The Contractor
   C. The Party awarding the contract
   D. The Tenderee

2. Which of the following statements on this project is **NOT** true?
   A. The total construction area is 24,240 square meters.
   B. The Contract Period totals 2 years.
   C. The contracting scope covers seven items.
   D. Construction drawings is excluded in this contract documents.

3. As for the Rights and Obligations of both Parties, which one is **NOT** true?

   A. Party A shall provide confirmed construction drawings and conduct technical clarification.

   B. Party A has the right to dismiss the employee of Party B directly if Party A consider such employee is not able to undertake the relevant work.

   C. If Party B finds inconformity between project standards and design drawings, Party B should inform the designer and Party A in writing.

   D. Party B shall be fully responsible for fixing and restoring the damages to the project items before the project acceptance.

4. "Force Majeure" includes the following causes **EXCEPT** _____?

   A. Earthquake

   B. Act of government

   C. Typhoon

   D. Construction site accident

5. What does "**quadruplicate**" mean in the last paragraph?

   A. Two copies

   B. Three copies

   C. Four copies

   D. Five copies

Exercise 2

Fill in the blanks with the words given below. Change the form where necessary.

arbitration	submit	sign	facade	engineer
prevail	observe	voluntary	drainage	stipulate

This contract is hereby signed by the two parties concerned on the principles of equality, _____, fairness and good faith. The contracting scope is a commercial building, including civil _____, inner decoration, _____, electronic, water supply and _____ etc., Party A and Party B should _____ the Rights and Obligations _____ in this contract. Any dispute arising from or in connection with this contract shall be settled through friendly negotiation between the parties, and in the event of the dispute can NOT be settled through negotiation, either party may _____ the dispute to the China International Economic and Trade _____ Commission. The contract becomes effective from the date of _____ and stamping by both parties. In the event of any conflict between the English and Chinese, the latter shall _____.

Exercise 3

Translate the following sentences into Chinese.

1. This contract is hereby signed by the two parties concerned according to the "Contract Law of the People's Republic of China" on the principles of equality, voluntariness, fairness and good faith.

2. In the event of Party A consider Party B is indeed unable to further fulfill its contract obligations, then Party A has the right to terminate the contract, and Party B should vacate the project site within two weeks after receiving Party A's notice.

3. Party B shall participate in the technical clarification on construction drawings and explanations of work procedures organized by Party A, and should develop the construction plan and the progress schedule for Party A to review.

4. Party B should make salary/wages payment to all construction staffs in time on a monthly basis, and will be fully responsible for any liabilities caused by any loss or damage to Party A as a result of Party B's failure to pay salaries and wages according to the requirements.

5. Party B is not liable for failure to perform the contractor's obligations if such failure is as a result of Force Majeure such as fire, flood, government forces and other Force Majeure factors.

课文译文

建设工程施工合同

发包方（甲方）：TM 国际投资公司

承包方（乙方）：中星建筑有限公司

本合同由如上列明的甲、乙双方按照《中华人民共和国合同法》《中华人民共和国建筑法》《建筑安装工程承包合同条例》以及国家相关法律法规的规定，遵循平等、自愿、公平和诚实信用的原则，于 2016 年 10 月 15 日在新加坡签订。

一、工程概况

1. 工程名称：现代商业大厦

2. 工程地点：新加坡

3. 工程内容：钢混 25 层；建筑面积 24 240 平方米

4. 工程立项批准文号：1617010023

5. 资金来源：外资

6. 开工日期：2017 年 3 月 1 日

    竣工日期：2019 年 3 月 1 日

7. 合同价款：本合同价款为人民币 100 000 000.00 元，为固定总价合同价款。

二、工程承包范围

1. 土建工程

2. 室内装饰工程

3. 室外装饰工程（门窗、屋面、台阶等）

4. 电气工程

5. 空调工程

6. 给排水及采暖工程

7. 消防工程

三、合同文件

组成本合同的文件包括：

1. 本合同协议书

2. 中标通知书

3. 投标书及其附件

4. 标准、规范及有关技术文件

5. 图纸

6. 工程量清单

7. 工程报价或预算书

四、甲方的权利与义务

1. 向乙方提供经确认的施工图纸或做法说明五份，并向乙方进行现场交底，向乙方提供施工所需要的水电设备，并说明使用注意事项。

2. 指派甲方驻工地代表，负责合同的履行。对工程质量、进度进行监督检查，办理验收、变更事宜。甲方的工地代表如果认为乙方的某雇员行为不当或散漫不羁，或甲方认为该雇员不能胜任，则乙方应立即进行撤换。

3. 如甲方认为乙方确已无能力继续履行合同的，则甲方有权解除合同，乙方必须在接到甲方书面通知后两周内撤离场地以便甲方继续施工。对乙方完成工程量的结算不作为撤离场地的条件，结算应在竣工一个月之内完成。

4. 甲方按照付款约定向乙方支付工程款。本工程付款方式：工程开工后 5 个工作日内甲方支付总工程款的 30%，室外门窗及幕墙工程验收合格后 5 个工作日内甲方支付总工程款的 20%，室内中间验收完成后 5 个工作日内甲方支付总工程款的 20%，工程完工验收合格后，并于乙方向甲方提交完整的竣工资料后 5 个工作日内支付工程总价款的 30%。

五、乙方的权利和义务

1. 参加甲方组织的施工图纸或做法说明的现场交底，拟订施工方案和进度计划，交由甲方审定。施工过程中，乙方按照甲方的有关指示及修改变更要求进行相应调整。

2. 指派乙方驻工地代表，负责合同的履行，按照要求组织施工，保质保量、按期完成施工任务。解决由乙方负责的各项事宜。乙方驻工地代表在本公司的工作时间不得少于90%工作日。

3. 根据甲方和设计方一致通过的设计图纸进行施工，遵从甲方工地代表和设计方的指示，如乙方发现在工程标准和设计图之间存在不一致时，应以书面形式通知甲方和设计方。

4. 从工程开始至工程竣工验收报告上所列明的日期之日止，乙方应全权负责整项工程。不管任何原因导致的工程的损坏，乙方负责修理还原，以使整个工程竣工时能完全依照合同要求以及甲方工地代表的指示，状况良好。

5. 乙方应与每位施工人员订立务工合同，办理合法务工证件并承担费用。乙方应每月按时向每位施工人员发放工资，如果乙方未按规定发放工资而给甲方带来的任何损失，则乙方承担完全赔偿责任。

六、不可抗力

如因不可抗力诸如火灾、水灾、政府强令措施以及其他不可抗力的原因致使乙方不能按期完成项目，乙方不负违约责任。但应在合理的时间内向甲方报告所发生的不可抗力并提供有关政府部门的证明文件。

七、争议处理

凡因执行本合约或有关本合同所发生的一切争执，双方应以友好方式协商解决；如果协商不能解决，应提交中国国际经济与贸易仲裁委员会，根据该会的仲裁规则进行仲裁。仲裁裁决是终局的，对双方都有约束力。

八、合同生效

本合同以中英文书写，一式四份，双方各持两份，两种文本具有同等效力。签字盖章后生效。如中英文相冲突的，以中文为准。

甲方： 　　　　　　　　　　　乙方：

签订日期： 　　　　　　　　　签订日期：

备注：由于篇幅所限，此合同仅为一份简化合同，真实合同则更精细复杂。

References

[1] 梅阳春，邹辉霞. 建设工程招投标及合同管理[M]. 2版. 武汉：武汉大学出版社，2012.

[2] 陈丹. 巧嘴英语商务谈[M]. 北京：北京邮电大学出版社，2006.

[3] CONTRACTUS. Contract of Construction[EB/OL]. (2011-12-07) http://blog.sina.com.cn/slblog_4dae264e0100vu8k.html

Unit 4　Project Management

Lead in

Identify the pictures below. Match them with the words in the box.

| construction crew | construction project | negotiation |

| | | |

Section A

Dialogue 1

Mr. Shaw is the representative from the Owner. Mr. Frank is the representative from the Contractor. They are discussing the progress of the project.

S = Mr. Shaw　　F = Mr. Frank

S: Ok. Gentlemen, the purpose of our meeting today is to review and discuss the progress of this project. Firstly, I would like you to explain what you have achieved for the past month, Mr. Frank?

F: Generally speaking, the construction activities have been going smoothly. But because of the recent **adverse** weather, we **revised** the schedule slightly.

S: Although the revision of the schedule sounds reasonable, it is a requirement of the contract that whenever you revise the schedule, you must **submit** the revised schedule to us for approval.

F: We don't think it is a significant **alternation**. If you want it, we have it right here for you.

S: Thank you. Anyway, I want to remind you that everything should be strictly in accordance with the contract documents. Now, the progress is two weeks behind schedule. So I want to know what measures you are taking to catch up with the schedule.

F: We are beginning to **implement** two-shift system for the **adit excavation** to **expedite** the progress.

S: Ok, as the representative of the project owner, I will, as always, stand firmly behind in the **execution** of the project.

F: Thank you. I believe that, through our **mutual** efforts, this project will surely be a success.

New Words

adverse	[ˈædvɜːs] adj.	not favorable; contrary 不利的；相反的
revise	[rɪˈvaɪz] v.	re-examine (sth), esp in order to correct or improve it 复查；（尤指）复核，校订，修正
submit	[səbˈmɪt] v.	give sth (to sb/sth) so that it may be considered, decided on, etc 呈递；提交
alternation	[ˌɔːltəˈneɪʃn] n.	successive change from one thing or state to another and back again 交互；轮流；间隔
implement	[ˈɪmplɪmənt] vt.	carry out or put into practice 使生效，执行
adit	[ˈædɪt] n.	a nearly horizontal passage from the surface into a mine 入口；平坑
excavation	[ˌekskəˈveɪʃn] n.	the act of digging 挖掘；发掘
expedite	[ˈekspədaɪt] vt.	speed up the progress of; facilitate 加速；加快；有助于

execution	[ˌeksɪˈkju:ʃn] n.	the carrying out, performance, or completion of an order, plan, or piece of work 实行，执行
mutual	[ˈmju:tʃuəl] adj.	having or based on the same relationship one towards the other 相互的，彼此的
		shared by two or more people 共同的，共有的

Phrases and Expressions

smoothly	进展顺利
submit to	顺从
in accordance with	与……一致，依照，根据
behind schedule	未按时完成，比原计划推迟
stand behind	做后盾，后援

Dialogue 2

Quality is one of the critical factors to ensure the success of a construction project. Due to the recent quality defects in the construction, a special meeting is called on the job site by the Owner's Representative, Mr. Ryan.

R = Mr. Ryan   B = Mr. Brown

R: What are you going to do with the problem?

B: Actually, we are doing the job strictly in **compliance** with the working drawings **issued** by your engineer. May I refer you to Drawing NO.MJ-9?

R: I will check the drawings and discuss the matter with our **consultants** this afternoon. Meanwhile, you are instructed to stop the work on this part of the project.

B: All right, Mr. Ryan. We take your instruction, but will you please give us a written confirmation?

R: It will be given to you soon after this meeting.

B: Well, Mr. Ryan. We will take a lesson from these matters and strengthen our quality **assurance** system?

R: I have to **reiterate** that **ongoing** construction work is at a critical stage. I expect your site management to **exert** close **supervision** and proper coordination of the various work **crews** without **compromising** quality.

B: It is true that we are facing a pressing completion and have to expedite the construction progress, but we will certainly not do it at the cost of project's quality. I believe that efforts on both sides will result in not only a timely but also a high-quality one.

New Words

compliance	[kəmˈplaɪəns] n.	action in accordance with a request or command; obedience 服从；听从；遵从
issue	[ˈɪʃuː] v.	come out of; prepare and issue for public distribution or sale 发行；发给；排出
consultant	[kənˈsʌltənt] n.	a person who gives specialist professional advice to others 顾问
assurance	[əˈʃʊərəns] n.	a firm statement that sth is certainly true or will certainly happen; promise 保证，确信
reiterate	[riˈɪtəreɪt] vt.	to say, state, or perform again 重申
ongoing	[ˈɒngəʊɪŋ] adj.	currently happening 进行的；不断发展的
exert	[ɪgˈzɜːt] v.	bring (a quality, skill, pressure, etc.) into use; make an effort 用(力)；尽(力)
supervision	[ˌsjuːpəˈvɪʒn] n.	management by overseeing the performance or operation of a person or group 监督；管理
crew	[kruː] n.	a group of people working together 一队（一班，一组）工作人员
compromising	[ˈkɒmprəmaɪzɪŋ] adj.	making or willing to make concessions 妥协的；损害名誉的；有失体面的

Phrases and Expressions

in compliance with	服从；听从；遵从；顺从
be instructed to	被指示，命令，通知做……
take one's instruction	接受（某人的）指示

Exercise 1

Decide whether the following statements are true (T) or false (F) according to the dialogues.

☐ 1. In Dialogue 1, the meeting aims to review and discuss the progress of the project.

☐ 2. According to Dialogue 1, the construction activities have been going smoothly on schedule.

☐ 3. In Dialogue 1, they are planning to implement two-shift system for bridging to catch up with the schedule.

☐ 4. In Dialogue 2, the contractor is required to exert strict supervision on the quality of the project.

☐ 5. In Dialogue 2, the project encountered a quality problem and some part of the project is required to stop.

Exercise 2

Oral practice.

Directions: Pair work. Use the questions below to interview your partner and then change roles.

1. When will you be able to submit the preliminary design to us?

_____

_____

2. We must point out that your actual progress is already one week behind schedule. Would you explain for that?

_____

_____

3. How about the progress at the power house?

_____

_____

4. Why have you reduced the progress?

_____

_____

5. Will the modification of the design affect the functions of the whole facility?

___

___

Exercise 3

Practical Activity.

Directions: Work in pairs and create a dialogue according to the given situation. In the project management, usually the Owner and the Contractor will convene a monthly progress review meeting. Representatives of each part will attend the meeting to review and discuss the previous month's progress. The result of the discussion will be performed to the whole class.

Section B

Project Management Processes

Project management is the discipline of **initiating**, planning, **executing**, controlling, and closing the work of a team to achieve specific goals and meet specific success criteria. A project is a temporary endeavor designed to produce a unique product, service or result with a defined beginning and end (usually **time-constrained**, and often constrained by funding or **deliverables**) undertaken to meet unique goals and objectives, typically to bring about beneficial change or added value. The **temporary** nature of projects stands in contrast to business as usual (or operations), which are repetitive, **permanent**, or semi-permanent functional activities to produce products or services. In practice, the management of these two systems is often quite different, and as such requires the development of distinct technical skills and management strategies. The primary challenge of project management is to achieve all of the project goals within the given constraints. This information is usually described in a user or project **manual**, which is created at the beginning of the development process. The primary constraints are scope, time, quality and **budget**. The secondary – and more **ambitious** – challenge is to **optimize** the **allocation** of necessary inputs and **integrate** them to meet pre-defined objectives.

Traditionally, project management includes a number of elements: four to five project management process groups. Regardless of the **methodology** or **terminology** used, the same basic project management processes or stages of development will be used. Major process groups generally include:

Initiating

The initiating processes determine the nature and scope of the project. If this stage is not performed well, it is unlikely that the project will be successful in meeting the business' needs. The key project controls needed here are an understanding of the business environment and making sure that all necessary controls are incorporated into the project. Any **deficiencies** should be reported and a **recommendation** should be made to fix them.

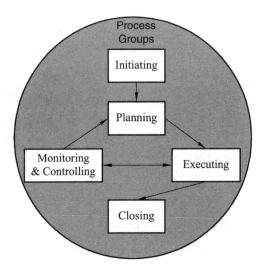

The initiating stage should include a plan that **encompasses** the following areas:
- Analyzing the business needs/requirements in measurable goals;
- Reviewing of the current operations;
- Financial analysis of the costs and benefits including a budget;
- Stakeholder analysis, including users, and support personnel for the project;
- Project charter including costs, tasks, deliverables, and schedules.

Planning

After the initiation stage, the project is planned to an appropriate level of detail. The main purpose is to plan time, cost and resources adequately to estimate the work needed and to effectively manage risk during project execution. As with the Initiation process group, a failure to adequately plan greatly reduces the project's chances of successfully accomplishing its goals.

Project planning generally consists of:
- Determining how to plan (e.g. by level of detail or Rolling Wave planning);
- Developing the scope statement;
- Selecting the planning team;
- Identifying deliverables and creating the work breakdown structure;
- Identifying the activities needed to complete those deliverables and networking the activities in their logical sequence;
- Estimating the resource requirements for the activities;

- Estimating time and cost for activities;
- Developing the schedule;
- Developing the budget;
- Risk planning;
- Gaining formal approval to begin work.

Executing

The execution/implementation phase ensures that the project management plan's deliverables are executed accordingly. This phase involves proper allocation, co-ordination and management of human resources and any other resources such as material and budgets. The output of this phase is the project deliverables.

Monitoring and controlling

Monitoring and controlling consists of those processes performed to observe project execution so that potential problems can be identified in a timely manner and corrective action can be taken, when necessary, to control the execution of the project. The key benefit is that project performance is observed and measured regularly to identify **variances** from the project management plan.

Monitoring and controlling includes:
- Measuring the ongoing project activities (where we are);
- Monitoring the project variables (cost, effort, scope, etc.) against the project management plan and the project performance baseline (where we should be);
- Identifying corrective actions to address issues and risks properly (How can we get on track again);
- Influencing the factors that could **circumvent** integrated change control so only approved changes are implemented.

In multi-phase projects, the monitoring and control process also provides feedback between project phases, in order to implement **corrective** or **preventive** actions to bring the project into compliance with the project management plan.

Project maintenance is an ongoing process, and it includes:
- Continuing support of end-users;
- Correction of errors;
- Updates to the product over time.

In this stage, auditors should pay attention to how effectively and quickly user problems are resolved.

Over the course of any construction project, the work scope may change. Change is a normal and expected part of the construction process. Changes can be the result of necessary design **modifications**, differing site conditions, material **availability**, contractor-requested changes, value engineering and impacts from third parties, to name a few. Beyond executing the change in the field, the change normally needs to be documented to show what was actually constructed. This is referred to as change management.

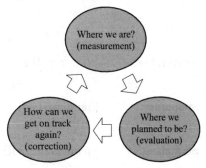

When changes are introduced to the project, the **viability** of the project has to be re-**assessed**. It is important not to lose sight of the initial goals and targets of the projects. When the changes **accumulate**, the forecasted result may not justify the original proposed investment in the project. Successful project management identifies these components, tracks and monitors progress so as to stay within time and budget frames already outlined at the commencement of the project.

Closing

Closing includes the formal acceptance of the project and the ending **thereof**. Administrative activities include the **archiving** of the files and documenting lessons learned.

This phase consists of:

Contract closure: Complete and settle each contract (including the resolution of any open items) and close each contract **applicable** to the project or project phase;

Project close: **Finalize** all activities across all of the process groups to formally close the project or a project phase;

Also included in this phase is the Post Implementation Review. This is a vital phase of the project for the project team to learn from experiences and apply to future projects. Normally a Post Implementation Review consists of looking at things that went well and analyzing things that went badly on the project to come up with lessons learned.

New Words

initiate	[ɪˈnɪʃɪeɪt] vt.	set (a scheme, etc.) working 开始，着手
execute	[ˈeksɪkjuːt] vt.	carry out; put into effect 执行；使生效
constrain	[kənˈstreɪn] vt.	hold back or force into an unwanted action, by limiting sb's freedom to act or choose 强迫，迫使；限制，约束
deliverable	[dɪˈlɪvərəbl] n.	something that can be provided as the product of development 应交付的产品
temporary	[ˈtemprəri] adj.	lasting only for a limited time 临时的，暂时的，短时间的
permanent	[ˈpɜːmənənt] adj.	lasting or intended to last for a long time or forever 永久（性）的，固定的；常置的，终身

manual	['mænjuəl] n.	book giving information about how to do sth 手册，指南
budget	['bʌdʒɪt] n. & v.	n. a plan of how to spend money 预算 v. arrange or plan (money, time, etc.) 编制预算，安排开支
ambitious	[æm'bɪʃəs] adj.	needing a lot of effort, money or time to succeed 费力的；耗资的；耗时的
optimize	['ɒptɪmaɪz] vt.	make optimal; get the most out of; use best 使完善；使优化
allocation	[ˌæləˈkeɪʃn] n.	the act of distributing by allotting or apportioning; distribution according to a plan 分配；配置；安置
integrate	['ɪntɪgreɪt] v.	join to sth else so as to form a whole （使）融入，（使）结合
update	[ˌʌp'deɪt] v.	bring (sth) up to date; modernize 更新（某事物），使现代化
modification	[ˌmɒdɪfɪ'keɪʃn] n.	n. the act or process of changing sth in order to improve it or make it more acceptable 修改，修饰
availability	[əˌveɪlə'bɪləti] n.	the quality of being at hand when needed 有效；有用；可得到的人（或物）；可用性
viability	[ˌvaɪə'bɪləti] n.	capable of being done in a practical and useful way 可行性
assess	[ə'ses] v.	decide or fix the value of (sth); evaluate 确定，评定（某事物）的价值；估价
accumulate	[ə'kju:mjəleɪt] v.	gradually get or gather together an increasing number or quantity of (sth) 积累，聚积（某物）
applicable	[ə'plɪkəbl] adj.	appropriate or suitable 可应用的，适当的
finalize	['faɪnəlaɪz] v.	put (sth) into final form; complete 使（某事物）达到最后形式；使完成

Phrases and Expressions

stand in contrast to	与……形成反差（对比）
regardless of	不顾，不惜
be unlikely to	不可能……
be incorporated to	被融入，纳入，合并到……
when necessary	在必要的时候

Exercise 1

Choose the best answer to each of the following questions.

1. According to paragraph one, which is correct about project management?
   A. Project management is a subject of organizing the work of a team to achieve specific goals and meet specific success criteria.
   B. Project management is a temporary endeavor designed to produce a unique product.
   C. Project management is repetitive, permanent, or semi-permanent functional activities to produce products or services.
   D. Project management and business management are similar.

2. What is the **NOT** the challenge of project management?
   A. To achieve all of the project goals within the given constraints.
   B. To optimize the allocation of necessary inputs to meet pre-defined objectives
   C. To integrate the necessary inputs to meet pre-defined objectives
   D. Scope, time, quality and budget.

3. Which of the following description of Initiating and Planning is correct?
   A. The planning processes determine the nature of the scope f the project.
   B. The project is planning to a more specific level in planning stage.
   C. The purpose of initiating is to plan time, cost and recourses adequately.
   D. Project planning consists of reviewing of the current operations.

4. Which of the following statements about monitoring and controlling is **NOT** true?
   A. The output of this phase is the project deliverables.
   B. Beyond executing the change in the field, the change normally needs to be documented to show what was actually constructed.
   C. The viability of the project has to be re-assessed when changes are introduced to the project.
   D. The good point of this stage is that project performance is observed and measured regularly to identify variances from the project management plan.

5. What is the theme of the passage?
   A. The methods to execute project management.

B. Introduction of the major project management process groups

C. Introduction of the discipline of project management

D. Introduction of the approaches of project management

Exercise 2

Fill in the blanks with the words given below. Change the form where necessary.

| modify | accumulate | ambitious | manual | temporary |
| integrate | deficiency | execute | available | budget |

1. Those who have entered for a temporary stay shall register for _____ residence in accordance with the same provisions.

2. Anyone who knew George would have described him as a hard-working, _____, competitive man with a splendid future in the company.

3. A workshop _____ gives diagrams and instructions for repairing your car.

4. U.S. forces are fully prepared for the _____ of any action once the order is given by the president.

5. If we _____ carefully, we'll be able to afford a new car.

6. We believe that students of _____ schools will have more tolerant attitudes.

7. The design of the spacecraft is undergoing extensive _____.

8. The shop has about 500 autographed copies of the book _____ for purchase.

9. Climate change is a result of _____ emission over many years, and is expected to be solved through long time of hard work as well.

10. Humans suffered as a consequence from nutritional _____ in almost all parts of the world because there weren't complete diets

Exercise 3

Translate the following sentences into Chinese.

1. A project is a temporary endeavor designed to produce a unique product, service or

result with a defined beginning and end.

_____

_____

2. The temporary nature of projects stands in contrast to business as usual (or operations), which are repetitive, permanent, or semi-permanent functional activities to produce products or services.

_____

_____

3. The execution/implementation phase ensures that the project management plan's deliverables are executed accordingly.

_____

_____

4. A failure to adequately plan greatly reduces the project's chances of successfully accomplishing its goals.

_____

_____

5. If this stage is not performed well, it is unlikely that the project will be successful in meeting the business' needs.

_____

_____

课文译文

项目管理

项目管理是启动、规划、执行、监控和结束团队工作以达到特定目标和实现特定成功标准的学科。一个项目需要在一定时期内为之努力，有明确的起止时间（通常有时间限制，也往往受资金和交付要求的约束），需要合理安排以生产出一个独特的产品、实现一种特殊的服务或完成一个特定的任务，通常会带来有益的变化或额外的价值。项目的临时性与普通商务（或业务）形成对比，后者是重复的、永久的或半永久的生产产品或服务的功能性活动。事实上，这两个系统的管理往往是完全不同的，因此需要发展不同的技术技能和管理策略。

项目管理的首要挑战是在给定的约束条件下实现所有的项目目标。此信息通常在用户或项目手册中描述，该手册在开发过程开始时创建。主要约束条件包括业务范围、时间、质量和预算。其次，也是更令人费心的挑战是，优化分配和整合必要的投入以满足预先设定的目标。

传统上，项目管理包括许多要素：四至五个项目管理过程组。无论使用哪种方法或术语，都将运用到一些相同的基本项目管理过程或开发阶段。主要的过程组一般包括：

1. 启动过程组

启动过程组确定项目的性质和范围。如果这个阶段执行不好，项目不可能成功地满足业务需求。这里需要的关键项目控制是了解业务环境，并确保所有必要的控制纳入到项目中。任何不足都应当报告指出，并提出改进建议。

启动阶段应具备一个包含以下方面的计划：

- 基于可衡量的目标分析业务需求；
- 回顾近期操作；
- 财务分析，包括成本和收益的预算；
- 利益相关者分析，包括用户和项目支持人员；
- 拟定项目章程，包括成本、任务、交付和时间表。

2. 规划过程组

启动阶段后，应从细节层面对该项目拟订详细计划。其主要目的是充分合理地安排时间、成本和资源，以预估所需的工作，并有效地解决项目执行过程中的风险。与启动过程组一样，未能充分计划会大大降低项目成功实现其目标的机会。

项目规划一般包括：

- 决定如何计划（例如通过细节或滚动式规划水平）；
- 制定范围说明；
- 选择规划团队；
- 确定可交付成果和创建任务分解结构；
- 确定完成可交付成果所需的活动并按照逻辑顺序组织活动；
- 估算活动资源需求；
- 估算活动的时间和成本；
- 制定进度表；

- 制定预算；
- 风险规划；
- 获得正式批准开始工作。

3. 执行过程组

执行过程组要确保项目的可交付成果按管理计划生产执行。这个阶段需要对人力资源和其他资源（如材料和预算）进行合理分配、协调和管理。此阶段的产出就是项目可交付成果。

4. 监控过程组

监控过程组包括那些观察项目执行的过程，以便及时发现潜在问题，并在必要时采取纠正措施以控制项目的各个过程组成。这个过程组的重要好处是观察并定期测量项目的绩效，以便识别项目管理计划在执行中的偏差。

监控过程组包括的过程有：
- 测量正在进行的项目活动（进行到哪个阶段）；
- 对照项目管理计划和项目绩效基准监控项目变量（成本、努力、范围等）（应当进行到哪个阶段）；
- 明确纠正措施以妥善解决问题和风险（如何才能再次走上正轨）；
- 对妨碍整体变更控制的因素施加影响，以做到仅实施经过批准的变更。

在多阶段项目中，监控过程组还在项目阶段之间提供反馈，以便实施纠正或预防措施，使项目符合项目管理计划的要求。

项目维护是一个持续的过程，它包括：
- 终端用户的持续支持；
- 错误的纠正；
- 随着时间推移产品的更新。

在这个阶段，审核员应该注意如何有效和快速地解决用户问题。

在任何建设项目的过程中，工作范围都可能发生变更。变更是施工过程中正常和预期之内的部分。变更可能源于必要的设计修改、不同的现场条件、材料的可用性、承包商要求的变化、价值工程和来自第三方的影响，仅举几例。除了在现场执行变更外，通常还需要记录下实际变更的内容，称为变更管理。

当项目引入变更时，项目的可行性必须重新评估。重要的是不要忘记项目的初始目标。当变化积累，预测的结果可能证明原来提出的投资项目不合理。成功的项目管理能识别出这些部分，跟踪和监控进度，以便保持在项目开始时所构建的时间和预算框架内。

5. 收尾过程组

收尾过程组包括对项目的正式验收及其完结。收尾管理活动包括文件归档和整理经验。

这个阶段包括：

合同收尾：完成和了结每个合同（包括任何开放项目的决议），并终止适用于项目或项目阶段的每个合同。

项目收尾：最终完成所有项目过程组的所有活动，正式结束项目或项目阶段。

包含在这个阶段的还有工作回顾。这是项目过程中的重要阶段，项目组从中总结经验并应用到未来的项目中。通常情况下，工作回顾包括审视和分析项目工作中好的和不足的部分以总结经验教训。

References

[1]　张水波,刘英. 国际工程管理实用英语口语[M]. 北京:中国建筑工业出版社, 1997.

Unit 5　Geotechnical Engineering

Identify the pictures below. Match them with the words in the box.

| deep foundation | landslide | piling |

Section A

Dialogue 1

Li Yang, an intern of a construction site, is going to inquire some information about piling. Peter is an experienced piler, who will give the answers to Li Yang.

L=Li Yang　　P=Peter

L: If you don't mind, I'd like to ask you some questions about **piling**, Peter.

P: Never mind.

L: What do you do before driving the piles?

P: There are three points. First, we make the site even and solid enough for freely piling machines.

L: What is then?

P: Then, we recheck the **bench marks** and the **axial** lines between the pile centers for determination of piles position.

L: I see. What about the last?

P: Finally, we find out and take away the barriers on the piling areas.

L: Obviously, you've gotten so much piling experience.

P: Yes, I've done the job for more than twenty years.

L: Oh, I've got some questions to ask.

P: OK, please go head.

L: Supposing the pile can't be driven further deep but the depth of piling is still far from design, what are you going to do?

P: It's a practical problem. In this case, we'll stop driving and study the driving records, and then make holes to find out the real situation under earth.

L: What will you do after you find out the real situation?

P: We can increase the pile length or **relocate** the pile position after consulting the designers.

L: I see.

New Words

piling	['paɪlɪŋ] n.	a column of wood or steel or concrete that is driven into the ground to provide support for a structure 打桩，打桩工程
axial	['æksɪəl] adj.	situated on or along or in the direction of an axis 轴的；成轴的
relocate	[ˌriːləʊˈkeɪt] vt.	become established in a new location 重新安置

Phrases and Expressions

bench marks   桩基准点

Dialogue 2

Will, a visitor of a site, is going to visit the underground works. Bruce is the worker of the site who will guide Will.

W= Will   B=Bruce

W: The quality of underground works is very critical for the whole buildings, how we couldn't expect such superstructures without very solid foundation? Could it be stable?

B: Absolutely true. High-rising buildings and large **mansions** never stand on any sand beach.

W: Right. What do you do next? I'm anxious to know.

B: It's our duty to keep the quality of the underground works complied with the **specification**.

W: I'm glad to hear that. By the way, what are your other workmates doing right now?

B: Most of them are piling. If time is available, let's go and watch, shall we?

W: That's a good idea. It's **gratified** for me to watch how to pile because I haven't specially arranged and freely used my time.

B: Ok. Let's go.

W: Thanks a million. You're most understanding.

B: You're welcome. Delighted I was able to help.

New Words

mansion	['mænʃn] n.	a large and imposing house 大厦；宅第；公馆
specification	[ˌspesɪfɪ'keɪʃn] n.	a detailed description of design criteria for a piece of work 详述规范；说明书
gratify	['grætɪfaɪ] vt.	make happy or satisfied 使高兴；使满意

Unit 5 Geotechnical Engineering 069

Exercise 1

Decide whether the following statements are true (T) or false (F) according to the dialogues.

☐ 1. "Never mind" here means that you can ask me any questions as you like.

☐ 2. Before driving the precast piles, we only make the site even and solid enough for moving piling machines.

☐ 3. When the pile can't be driven further deep but the depth of piling is still far from the design, then the real situation under earth have to be found out.

☐ 4. Before consulting the designers, we hardly make any decision to increase pole length or to relocate the pile position.

☐ 5. "Thanks a million" means no doubt at all.

Exercise 2

Oral practice.

Directions: Pair work. Use the questions below to interview your partner and then change roles.

1. What is the first point before piling?

_____

_____

2. What is the last point before driving the piles?

_____

_____

3. How long has the man done the job of piling?

_____

_____

4. What will you do if the barrier is found?

___

___

5. How do you guarantee the quality of underground works?

___

___

Exercise 3

Practical Activity.

Directions: Work in pairs. Mr. Li acts as a piler. Mr. Wang acts as a visitor. Try to talk with your partners about what you should pay attention to before and when you're piling. The result of the discussion will be reported to the whole class.

Section B

Geotechnical engineering

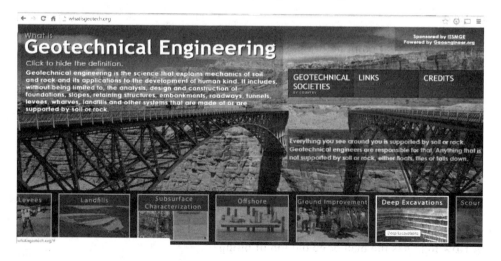

Geotechnical engineering is the branch of civil engineering concerned with the engineering behavior of earth materials. Geotechnical engineering is important in civil engineering, but also has applications in military, mining, **petroleum** and other engineering disciplines concerned with construction occurring on the surface or within the ground. Geotechnical engineering uses principles of soil mechanics and rock **mechanics** to investigate subsurface conditions and materials; determine the relevant physical/mechanical and chemical properties of these materials; evaluate stability of natural **slopes** and man-made soil deposits; assess risks posed by site conditions; design **earthworks** and structure foundations; and monitor site conditions, earthwork and foundation construction.

In the 19th century Henry Darcy developed what is now known as Darcy's Law describing the flow of **fluids** in **porous** media. Joseph Boussinesq (a mathematician and physicist) developed theories of stress distribution in **elastic** solids that proved useful for estimating stresses at depth in the ground; William Rankine, an engineer and physicist, developed an **alternative** to Coulomb's earth pressure theory. Albert Atterberg developed the clay consistency indices that are still used today for soil classification. Osborne Reynolds recognized in 1885 that shearing causes **volumetric** dilation of dense and contraction of loose **granular** materials.

Modern geotechnical engineering is said to have begun in 1925 with the publication of Erdbaumechanik by Karl Terzaghi (a civil engineer and geologist). Considered by many to be the father of modern soil mechanics and geotechnical engineering, Terzaghi developed the principle of effective stress, and demonstrated that the shear strength of soil is controlled by effective stress.

In geotechnical engineering, soils are considered a three-phase material composed of: rock or mineral particles, water and air. The **voids** of a soil, the spaces in between mineral particles, contain the water and air.

The engineering properties of soils are affected by four main factors: the **predominant** size of the mineral particles, the type of mineral particles, the grain size distribution, and the relative quantities of mineral, water and air present in the soil **matrix**. Fine particles (fines) are defined as particles less than 0.075 mm in diameter.

A variety of soil samplers exist to meet the needs of different engineering projects. The standard **penetration** test (SPT), which uses a thick-walled split spoon sampler, is the most common way to collect disturbed samples. Piston samplers, employing a thin-walled tube, are most commonly used for the collection of less disturbed samples. Properties such as shear strength, **stiffness hydraulic** conductivity, and **coefficient** of consolidation may be significantly altered by sample **disturbance**. To measure these properties in the laboratory, high quality sampling is required. Common tests to measure the strength and stiffness include the triaxial shear and unconfined compression test.

A building's foundation **transmits** loads from buildings and other structures to the earth. Geotechnical engineers design foundations based on the load characteristics of the structure and the properties of the soils and/or **bedrock** at the site. In general, geotechnical engineers: (a) Estimate the magnitude and location of the loads to be supported; (b) Develop an investigation plan to explore the subsurface; (c) Determine necessary soil parameters through field and lab testing (e.g., consolidation test, triaxial shear test, vane shear test, standard penetration test); (d) Design the foundation in the safest and most economical manner. In areas of shallow bedrock, most foundations may bear directly on bedrock; in other areas, the soil may provide sufficient strength for the support of structures. In areas of deeper bedrock with soft overlying soils, deep foundations are used to support structures directly on the bedrock; in areas where bedrock is not economically available, stiff "bearing layers" are used to support deep foundations instead.

Gravity walls depend on the size and weight of the wall mass to resist pressures from behind. Gravity walls will often have a slight setback, or batter, to improve wall stability. Prior to the introduction of modern reinforced-soil gravity walls, **cantilevered** walls were the most common type of taller retaining wall. Cantilevered walls are made from a relatively thin stem of steel-reinforced, cast-in-place concrete or mortared **masonry** (often in the shape of an inverted T).

Example of a slab-on-grade foundation

Shallow foundations are a type of foundation that transfers building load to the very near the surface, rather than a subsurface layer. Shallow foundations typically have a depth to width ratio of less than 1. Deep foundations are used for structures or heavy loads when shallow foundations cannot provide adequate capacity, due to size and structural limitations. They may also be used to transfer building loads past weak or compressible soil layers. While shallow foundations rely solely on the bearing capacity of the soil beneath them, deep foundations can rely on end bearing resistance, frictional resistance along their length, or both in developing the required capacity.

Footings (often called 'spread footings' because they spread the load) are structural elements which transfer structure loads to the ground by direct areal contact. Footings can be isolated footings for point or column loads, or strip footings for wall or other long (line) loads. Footings are normally constructed from reinforced concrete cast directly onto the soil, and are typically embedded into the ground to penetrate through the zone of frost movement and/or to obtain additional bearing capacity.

Slope stability is the analysis of soil covered slopes and its potential to undergo movement. Stability is determined by the balance of shear stress and shear strength. A previously stable slope may be initially affected by preparatory factors, making the slope conditionally unstable. Triggering factors of a slope failure can be climatic events, can then make a slope actively unstable, leading to mass movements. Mass movements can be caused by increases in shear stress, such as loading, lateral pressure, and transient forces. Alternatively, shear strength may be decreased by weathering, changes in pore water pressure, and organic material.

New Words

petroleum	[pə'trəʊlɪəm] n.	a dark oil consisting mainly of hydrocarbons 石油
mechanics	[mɪ'kænɪks] n.	the branch of physics concerned with the motion of bodies in a frame of reference 力学；机械学
slope	[sləʊp] n.	the property possessed by a line or surface that departs from the horizontal 斜坡

earthwork	[ˈɜːθwɜːk] n.	an earthen rampart 土方（工程）
fluid	[ˈfluːɪd] n.	a liquid or a gas 液体流体
porous	[ˈpɔːrəs] adj.	full of pores or vessels or holes; able to absorb fluids 孔渗水的；能渗透的
elastic	[ɪˈlæstɪk] adj.	able to adjust readily to different conditions 弹性的
alternative	[ɔːlˈtɜːnətɪv] adj.	serving or used in place of another 替代的；备选的；其他的
volumetric	[ˌvɒljʊˈmetrɪk] adj.	of or relating to measurement by volume [物]体积的；容积的
granular	[ˈɡrænjələ(r)] adj.	composed of or covered with particles resembling meal in texture or consistency 颗粒状的
void	[vɔɪd] adj.	containing nothing 无效的；空的
predominant	[prɪˈdɒmɪnənt] adj.	having superior power and influence 占主导地位的；占优势的
matrix	[ˈmeɪtrɪks] n.	(mathematics) a rectangular array of quantities or expressions set out by rows and columns [数]矩阵；模型
penetration	[ˌpenɪˈtreɪʃn] n.	the act of entering into or through something 渗透
stiffness	[ˈstɪfnəs] n.	the physical property of being inflexible and hard to bend 僵硬；坚硬
hydraulic	[haɪˈdrɔːlɪk] adj.	of or relating to the study of hydraulics 液压的；水力的
coefficient	[ˌkəʊɪˈfɪʃnt] n.	a constant number that serves as a measure of some property or characteristic [数]系数；率；协同因素
disturbance	[dɪˈstɜːbəns] n.	the act of disturbing something or someone 困扰
transmit	[trænsˈmɪt] vt.	send from one person or place to another 传送
bedrock	[ˈbedrɒk] n.	solid unweathered rock lying beneath surface deposits of soil 基岩；牢固基础
cantilever	[ˈkæntɪliːvə(r)] n.	projecting horizontal beam fixed at one end only 悬臂

Phrases and Expressions

geotechnical engineering 岩土工程

soil mechanics 土力学

rock mechanics 岩石力学

three-phase material 三相材料

shear strength 剪切强度

standard penetration test 标准贯入试验

building foundation 建筑基础

bearing layers 持力层

retaining wall 挡土墙

deep foundations 深基础

triggering factors 触发因素

cantilevered walls 悬臂式挡土墙

shallow foundations 浅基础

Exercise 1

Choose the best answer to each of the following questions.

1. Which of the following statement concerning geotechnical engineering is **NOT** true?

    A. It investigates subsurface conditions and materials.

    B. It determines the relevant mechanical and chemical properties.

    C. It evaluates risks posed by site conditions.

    D. It is considered a three-phase material composed of: rock or mineral particles.

2. Who developed theories of stress distribution in elastic solids that proved useful for evaluating stress at depth in the earth?

    A. Henry Darcy.

    B. Joseph Boussinesq.

    C. William Rankine.

    D. Karl Terzaghi.

3. What relies on the size and weight of the wall mass to resist stress from behind?

    A. Gravity walls.

    B. Cantilevered walls.

    C. Shallow foundations.

    D. Deep foundations.

4. What is the purpose of the slope stability?

    A. It determines the balance of shear strength.

B. It is affected by preparatory factors.

C. It analyzes the soil covered slopes and its potential to undergo movement.

D. It analyzes the soil's potential to move.

5. Which of the following statement is true?

A. Shallow foundations are used to support structures directly on the bedrock.

B. Gravity walls are the most common type of retaining wall.

C. Terzaghi developed the clay consistency indices which are used for soil classification.

D. Geotechnical engineering is vital in civil engineering and also applies in other disciplines such as mining and military.

Exercise 2

Fill in the blanks with the words given below. Change the form where necessary.

alternative	geotechnical	transmit	stiffness	void
predominant	disturbance	cantilever	mechanic	elastic

1. The information is electronically _____ to schools and colleges

2. The environmental effect of underground construction include those on _____ problems, water supply, water chemistry and vegetation.

3. The _____ of demand for a single newspaper is bound to be higher than the figure for newspapers as a whole.

4. The Supreme Court threw out the confession and _____ his conviction for murder.

5. His group was forced to turn back and take an _____ route.

6. The furniture was _____, uncomfortable, too delicate, and too neat

7. He has not studied _____ engineering.

8. According to our characteristic of enterprise to run a college, it discusses the _____ operation mechanism to implement the "1+ N" evaluating pattern.

9. Find a quiet, warm, comfortable room where you won't be _____.

10. A high-rise building is nothing but a slender, vertical, _____ beam resisting lateral and vertical loads.

Exercise 3

Translate the following sentences into Chinese.

1. A geotechnical engineer determines and designs the type of foundations, earthworks, and pavement subgrades required for the intended man-made structures to be built.

_____

_____

2. Foundations are bases, usually of concrete, placed under the ground so as to spread a vertical load over it.

_____

_____

3. Excavation through soft or granular soils requires casing.

_____

_____

4. The basic concept in the base isolation technique is to introduce flexibility at the base of structural frames in the horizontal direction.

_____

_____

5. Stability analysis is needed for the design of engineered slopes and for estimating the risk of slope failure in natural or designed slopes.

_____

_____

课文译文

岩土工程

岩土工程是土木工程的一门分支学科，以岩土体作为研究对象。岩土工程不光应用于土木工程，还应用于军事工程、采矿工程、石油工程，及其他与地上、地下建造活动有关的工程。岩土工程使用土力学和岩石力学的原则研究地下状况和材料，确定这些材料的相关物理/力学和化学性质，预估天然斜坡的稳定性和人工的土壤沉积，评估现场风险，设计土方和结构基础，并监测现场、土方和基础施工。

早在 19 世纪的时候，亨利·达西就提出了被现代人熟知的达西定律，描述了水在多孔隙介质中的渗流规律。约瑟夫·布西涅斯克（著名数学家、物理学家）提出了弹性固体中应力分布理论，后被应用于估计土体的应力分布。威廉·朗金（著名工程师、物理学家）提出了可替代库仑土压力理论的朗金土压力理论。艾伯特·阿太堡提出了土的塑性指数，用于土的分类，一直沿用至今。1885 年，奥斯本·雷诺发现了剪切力会引起土体的体积膨胀和压缩。

现代岩土工程学开始于 1925 年，其标志是凯尔·太沙基（著名土木工程师、土力学家）发表了他的土力学专著《建立在土的物理学基础的土力学》。凯尔·太沙基也被誉为是现代土力学、现代岩土工程学之父，他提出的土的有效应力原理表明了土的抗剪强度取决于土的有效应力。

在岩土工程中，土是一种三相材料，其成分包括岩石或矿物颗粒、水和空气。在土的孔隙里，即矿物颗粒之间，包含着水和空气。

土的工程性质主要受四个因素影响：矿物颗粒的主要尺寸与类型，颗粒级配，矿物质的相对含量，土壤中的水和空气含量。细颗粒的定义是粒径小于 0.075 毫米的颗粒。

不同的工程项目需要使用不同的土壤采样器。标准贯入试验（SPT）使用的是一个厚壁对开式取土器，这是最常见的扰动土样取土器。活塞取土器采用的是薄壁管，是最常用的不扰动土样取土器。

对于扰动土样来说，土壤的特性如抗剪强度、渗透系数，以及压实系数都会发生很明显的改变。为了能在实验室中准确测试土壤的这些特性，就需要对土壤进行高质量的采样。常见的测试土体强度和硬度的试验有三轴剪切试验和无侧限抗压强度试验。

工程师设计建筑基础时是根据建筑结构荷载类型和现场土壤、基岩性质设计的。一般岩土工程师会做以下几方面工作：（a）估计基础将承受的荷载大小及负载位置；（b）制订工程地质勘察计划；（c）通过现场测试和实验室确定必要的土壤参数（例如，固结试验、三轴剪切试验、十字板剪切试验、标准贯入试验）；（d）以最安全、最经济的方式设计建筑基础。在浅层基岩地区，大多数基础设计是直接让基岩承压；但在其他一些地区，土壤层能够提供足够的强度承受建筑物荷载。当在深层基岩地区且覆盖层为软土时，常常采用深基础的方式让基岩承压；如果采用深基础的方式在经济上不可行，可利用坚硬的"持力层"代替基岩承受深基础传递下来的荷载。

重力式挡土墙依靠自身墙体的体积和重量抵抗背后的土压力。重力式挡土墙通常有一点仰斜或俯斜，这样能提高墙体的稳定性。在引入现代的加筋土挡土墙之前，悬臂式挡土墙是最常见的高挡土墙类型。悬臂式挡土墙采用较细的钢筋，现场浇筑混凝土或砌筑砂浆砌块完成(通常设计为倒 T 的形状)。

浅基础是一种将建筑荷载转移到非常接近表面的地基，而不是地下层的基础类型。浅基础通常具有小于 1 的深宽比。当浅基础由于尺寸和结构限制而不能提供足够的能力时，深基础用于结构或重载荷。它们还可以用于将建筑物荷载传递经过弱的或可压缩的土层。浅基础仅依赖于它们下面的土壤的承载能力；而深基础可以依赖于端部轴承阻力、沿着它们的长度的摩擦阻力，或两者同时发挥作用。

基础（通常称为"扩展基础"，因为它们扩展了载荷）是通过直接面接触将结构载荷转移到地面的结构元件。基础可以是用于点或柱负载的隔离基础，或用于墙壁或其他长（线）负载的条带基础。基础通常由直接浇注到土壤上的钢筋混凝土构成，并且通常嵌入地下以穿透霜运动区域和/或获得额外的承载能力。

边坡稳定性是土壤覆盖的斜坡承受和接受的可能性运动。稳定性取决于剪应力和剪切强度的平衡。先前稳定的斜率可能最初受到准备因素的影响，使得斜率有条件地不稳定。触发因素，边坡失稳可再气候事件可以使一个斜坡积极不稳定，导致块状运动。质量运动可以由剪切应力的增加引起,例如负载、侧向压力和瞬时力。同样地,风化作用、孔隙水压及有机材料的改变可能导致剪切强度的减小。

References

[1] Terzaghi K, Peck RB, Mesri G. Soil Mechanics in Engineering Practice 3rd Ed. John Wiley & Sons, Inc, 1996.

[2] Holtz R, Kovacs W. An Introduction to Geotechnical Engineering. Prentice-Hall, Inc, 1981.

[3] Das, Braja. Principles of Geotechnical Engineering. Thomson Learning, 2006.

[4] Budhu, Muni. Soil Mechanics and Foundations. John Wiley & Sons, Inc, 2007.

Unit 6 Highway Engineering

Lead in

Identify the pictures below. Match them with the words in the box.

| construction | freeway | maintenance |

Section A

Dialogue 1

Mr. Wilson, a representative of the owner, is talking about the project progress with Mr. Zhang.

W=Mr. Wilson   Z=Mr. Zhang

Z: Welcome to the construction site. I'm the project manager.

W: Nice to meet you! How is the project going?

Z: Fairly well. As you can see, we have finished the **removal** of earth and old buildings ahead of schedule. More pieces of equipment and building materials are required for **excavation** and **pavement**. A few days ago, we purchased some **excavators**, **bulldozers** and **scrapers**.

W: That's great! How about the material supply?

Z: We have ordered 500 tons of Portland **cement**, 300 tons of **gravel** and 350 tons of steel. Due to the bad weather, these materials have not been **delivered** to the job site.

W: Will the project be **postponed**?

Z: As soon as these materials are delivered, a two-shift system will be applied immediately to **expedite** the project progress.

W: In other words, smooth progress of the project will be **guaranteed**.

Z: Exactly!

New Words

removal	[rɪˈmuːvl] n.	the act of removing 清除
excavation	[ˌekskəˈveɪʃn] n.	the act of digging 挖掘
pavement	[ˈpeɪvmənt] n.	the paved surface of a thoroughfare 路面
excavator	[ˈekskəveɪtə(r)] n.	a machine for excavating 挖掘机
bulldozer	[ˈbʊldəʊzə(r)] n.	large powerful tractor for moving earth 推土机
scraper	[ˈskreɪpə(r)] n.	machine that can be used for scraping a particular surface clean 铲土机
cement	[sɪˈment] n.	a building material that is a powder made of a mixture of limestone and clay 水泥
gravel	[ˈɡrævl] n.	rock fragments and pebbles 碎石;砂砾
deliver	[dɪˈlɪvə(r)] vt. & vi.	bring to a destination; make a delivery 递送;传送
postpone	[pəˈspəʊn] vt. & vi.	hold back to a later time 延期
expedite	[ˈekspədaɪt] vt.	speed up the progress of; facilitate 加快;促进
guarantee	[ˌɡærənˈtiː] vt.	promise to do or accomplish 保证

Phrases and Expressions

project progress 项目进度
project manager 项目经理
ahead of schedule 提前
Portland cement 波特兰水泥

Dialogue 2

Mr. Zhang, Mr. Wilson and Mr. Johnson are discussing environmental protection at the meeting.

Z=Mr. Zhang　　　W=Mr. Wilson　　　J=Mr. Johnson

Z: Good morning, everyone!

W: Good morning, Mr. Zhang! Allow me to introduce Mr. Johnson, the **inspector** from the Environmental Protection Agency.

Z: I'm pleased to meet you, Mr. Johnson. I'm the project manager from China Construction Eighth Engineering Division Company limited. I'm in charge of the Shanghai-Chengdu **Freeway** Project.

J: The pleasure is mine, Mr. Zhang.

W: Mr. Johnson wants to know how the environment is protected while the project is going on.

J: Let me put it this way, many residents are rather concerned with **adverse** environment impact of freeway construction. Have you taken some measures?

Z: Sure! We have studied environmental protection **regulations** published by the government and adopted **corresponding** measures.

J: May I have some specific details?

Z: For example, various **facilities** have been constructed for special treatment of waste water so as to prevent **contamination**.

J: That sounds fine.

Z: Let's go and look around, shall we?

J: All right.

New Words

inspector	[ɪnˈspektə(r)] n.	an investigator who observes carefully 检查员
freeway	[ˈfriːweɪ] n.	a broad highway designed for high-speed traffic 高速公路
adverse	[ˈædvɜːs] adj.	contrary to your interests or welfare 不利的
regulation	[ˌreɡjuˈleɪʃn] n.	a principle or condition that customarily governs behavior 规则
corresponding	[ˌkɒrəˈspɒndɪŋ] adj.	Accompanying; conforming in every respect 相应的
facility	[fəˈsɪləti] n.	equipment or services that are provided for a particular purpose 设施；设备
contamination	[kənˌtæmɪˈneɪʃən] n.	the state of being contaminated 污染

Phrases and Expressions

Environmental Protection Agency 环境保护局

China Construction Eighth Engineering Division Company Limited 中国建筑第八工程局有限公司

in charge of 负责；主管

put it this way 这样说；这样表达

be concerned with 关心

Exercise 1

Decide whether the following statements are true (T) or false (F) according to the dialogues.

☐ 1. The old buildings have been removed before excavation.

☐ 2. Building materials have been delivered to the job site.

☐ 3. The whole project seems to be postponed because of the bad weather.

☐ 4. The environmental impact of constructing freeways has received increased attention.

☐ 5. The freeway construction had negative effect on environment and high walls were built to prevent air pollution.

Exercise 2

Oral practice.

Directions: Pair work. Use the questions below to interview your partner and then change roles.

1. What are needed for excavation and pavement of the project?

_____

_____

2. Why is the delivery of building materials delayed?

_____

_____

3. How do the employees guarantee the smooth progress of the project?

_____

_____

4. What does Mr. Zhang do for a living?

_____

_____

5. What effective measures have been adopted to reduce the adverse impact of project construction on environment?

_____

_____

Exercise 3

Practical Activity.

Directions: Work in pairs. Student A and Student B are talking about a project. Student

A is the representative of the owner; student B is the project manager. Then switch the roles. The result of the discussion will be reported to the whole class.

Section B

Highway Engineering

A road is the way between two places, which allows the traveling by foot or a means of **conveyances**, including a horse, cart, or motorized vehicle. The beginning of road construction dated back to Romans times. Roman roads were characterized by their **linearity** and **durability** out of military factor. Normally, the surface of the roads was often elevated a meter or more above the local ground level so as to provide a clear view of surrounding regions for Romans. In this way, the modern term "**highway**" came into being.

Road generally consists of highway, urban road, factories and mines road, forest road, and country road. Highway is primarily applied to motor transport among different cities. Gradually, the advancement of technology and growth of economic activities among different countries contribute to the development of modern highway. The construction of a high-quality highway network directly increases a nation's economic output and improves people's living standard by reducing journey time and cost. Highway engineering, a discipline branching from civil engineering, involves highway planning, highway design, highway construction and highway **maintenance**.

Before the new highway is designed and constructed, a general planning and **evaluation** of traffic, financing, and environment must be considered. Highway engineers are responsible for assessing traffic needs of the area for a certain period, generally 20 years, by collecting information about **volume**, **distribution**, and characteristics of present traffic. By analyzing the traffic data and possible changes to be expected in these factors, highway engineers decide what construction will meet traffic needs.

Financing consideration determines whether the project can be carried out at one time or in different stages. The **legislation** has been passed by the state to encourage individual and corporate investment to speed up the development of highway construction. Engineers select the best construction plan by analyzing economic **feasibility** of the highway. Generally, users of highway and owners of **adjacent** property benefit from decreased cost of transportation, improved street parking and increased property value.

Nowadays, residents attach great importance to environmental impacts of highway construction. The adverse impacts of improperly planed and constructed highway commonly include noise **generation**, air pollution, water contamination, and damage of natural landscape. Many construction projects have been delayed or even canceled because of environmental problems. Therefore, Environmental Impact Assessment (EIA) is needed for highway engineers to identify and evaluate all the possible impacts of project construction on natural environment.

Detailed design of highway project includes **location**, pavement design and **drainage** design. After highway project planning is addressed, engineers set out to select the most appropriate location of the highway. To begin with, engineers take an overall look at the area concerned. By drawing **longitudinal** sections along possible **alignment**, engineers select the possible routes, all of which satisfy the design standards. The locator is then applied to explore the area so as to search out feasible routes and locate major **obstacles**. In this way, the most suitable route is finally decided by carefully examining each of possible routes.

Pavement or surface refers to the structure on which the durable surface materials laid down so as to sustain passenger flow. In the past, **cobblestones** and gravel highway surfaces are extensively used, but these surfaces have mostly been replaced by **asphalt** or **concrete**. Basically, the pavement structure consists of surface course, base course and **subbase** course. All the hard surfaced pavements are typically categorized into **flexible** pavement and **rigid** pavement. Flexible pavement is surfaced with asphalt materials. The word "flexible" is applied since the pavement structure **bends** due to traffic loads. Rigid pavement is surfaced with Portland Cement Concrete (PCC) and substantially stiffer than flexible pavement due to PCC's high stiffness. The **performance** of the pavement depends heavily on the satisfactory character of each component, which requires highway engineers' proper evaluation.

Drainage refers to the removal of surface and sub-surface water from an area. By building the road with a crown and sloping its shoulders and adjacent areas, surface drainage structure thus makes water flow either toward existing natural drainage structure or into a storm drainage system, such as catch basins and underground pipes. If the storm drainage system is applied, highway engineers should consider the maximum rate of **runoff** expected, duration of the storm, and the amount of each catch basin. When facilities are designed for removing sub-surface water under the highway, engineers must analyze the maximum probable **precipitation** and the highest estimated flow rate of the area, and then calculate the required capacity of drainage structure.

Highway construction processes mainly include excavation, **filling**, compacting, and **installation**. Firstly, old road surfaces, buildings and topsoil are removed. The filling is then made by the compacted layer method. In accordance with **specifications**, each layer of filling is spread and compacted. **Pneumatic** tired **rollers** are often used to compact the soil below the highway so as to **eliminate settlement**. The highway construction is completed by installing crash barriers, traffic signs, and other forms of road surface marking.

Highway maintenance consists of the repair and **upkeep** of pavements, drainage facilities, and traffic control devices. Highway deterioration is primarily due to **accumulated** damage from vehicles and environmental effect, such as frost heaves and **thermal** cracking. Highway is designed for an expected service life. **Virtually** all highways require some forms of maintenance before they come to the end of their service life. The relevant agencies continually apply preventive maintenance treatments to **prolong** the life **span** of highways.

New Words

conveyance	[kənˈveɪəns] n.	something that serves as a means of transportation 交通工具
linearity	[ˌlɪnɪˈærəti] n.	the property of having one dimension 直线性
durability	[ˌdjʊərəˈbɪləti] n.	permanence by virtue of the power to resist stress or force 耐久性
highway	[ˈhaɪweɪ] n.	a major road for any form of motor transport 公路
maintenance	[ˈmeɪntənəns] n.	activity involved in maintaining something in good working order 维护；维修
evaluation	[ɪˌvæljʊˈeɪʃn] n.	act of ascertaining or fixing the value or worth of 评价；评估
volume	[ˈvɒljuːm] n.	the amount of three-dimensional space occupied by an object 量；容积
distribution	[ˌdɪstrɪˈbjuːʃn] n.	an arrangement of values of a variable showing their frequency of occurrence 分布
legislation	[ˌledʒɪsˈleɪʃn] n.	law enacted by a legislative body 立法；法律
feasibility	[ˌfiːzəˈbɪləti] n.	the quality of being doable 可行性；可能性
adjacent	[əˈdʒeɪsnt] adj.	nearest in space or position 邻近的；毗连的
generation	[ˌdʒenəˈreɪʃn] n.	a coming into being 产生
location	[ləʊˈkeɪʃn] n.	the act of putting something in a certain place 定位
drainage	[ˈdreɪnɪdʒ] n.	emptying something accomplished by allowing liquid to run out of it 排水
longitudinal	[ˌlɒŋɡɪˈtjuːdɪnl] adj.	of or relating to lines of longitude 纵向的
alignment	[əˈlaɪnmənt] n.	the spatial property possessed by an arrangement of things in a straight line 线性

obstacle	[ˈɒbstəkl] n.	something immaterial that stands in the way and must be circumvented or surmounted 障碍
cobblestone	[ˈkɒbləstəun] n.	rectangular paving stone with curved top, once used to make roads 鹅卵石
asphalt	[ˈæsfælt] n.	a dark bituminous substance in natural beds and as residue from petroleum distillation 沥青
concrete	[ˈkɒŋkriːt] n.	a strong hard building material composed of sand and gravel and cement and water 混凝土
subbase	[ˈsʌbˌbeɪs] n.	the lowest molding of an architectural base or of a baseboard 垫层
flexible	[ˈfleksəbl] adj.	able to bend easily 易弯曲的
rigid	[ˈrɪdʒɪd] adj.	incapable of or resistant to bending 坚硬的
bend	[bend] vt. & vi.	form a curve 弯曲
performance	[pəˈfɔːməns] n.	process or manner of functioning or operating 性能
runoff	[ˈrʌnˌɔːf] n.	the occurrence of surplus liquid (such as water) exceeding the limit or capacity 径流
precipitation	[prɪˌsɪpɪˈteɪʃn] n.	the quantity of water falling to earth at a specific place within a specified period of time 降水
filling	[ˈfɪlɪŋ] n.	the act of filling something 填筑，填充
installation	[ˌɪnstəˈleɪʃn] n.	the act of installing something 安装
specification	[ˌspesɪfɪˈkeɪʃn] n.	a detailed description of design criteria for a piece of work 规格,说明
pneumatic	[njuːˈmætɪk] adj.	of or relating to or using air or a similar gas 充气的
roller	[ˈrəʊlə(r)] n.	a road leveling machine 压路机
eliminate	[ɪˈlɪmɪneɪt] vt.	remove something completely 消除
settlement	[ˈsetlmənt] n.	the process of formation of sedimentary rocks 沉降
upkeep	[ˈʌpkiːp] n.	activity involved in maintaining something in good working order 保养，维护
accumulate	[əˈkjuːmjəleɪt] vt. & vi.	collect or gather 累积，积聚
thermal	[ˈθɜːml] adj. & n.	relating to or associated with heat 热的 rising current of warm air 热气流
virtually	[ˈvɜːtʃuəli] adv.	almost, nearly 几乎；差不多
prolong	[prəˈlɒŋ] vt.	cause to be or last long 延长；拖延
span	[spæn] n. & v.	the distance between two points 跨度 to extend over an area or time period 跨越

Phrases and Expressions

date back to 追溯至
come into being 形成
apply to 应用于
contribute to 有助于；促成
highway construction 公路施工
highway maintenance 公路维护
carry out 执行；实行
speed up 加速
benefit from 得益于
attach importance to 重视
Environmental Impact Assessment 环境影响评估
set out to 开始；着手
surface course 面层
base course 基层
subbase course 垫层
flexible pavement 柔性路面
rigid pavement 刚性路面
Portland Cement Concrete 普通水泥混凝土；硅酸盐水泥混凝土
catch basin 集水箱，集水池
compacted layer method 分层填筑压实方法
in accordance with 根据；依照
crash barrier 防撞栏
road surface marking 路面标记
frost heave 冻胀
thermal cracking 热裂解

Exercise 1

Choose the best answer to each of the following questions.

1. What are the features of Roman roads?
   A. Linearity and durability
   B. Its long history

C. Economic feasibility

D. Detailed design of excavation

2. What should highway engineers consider before the design and construction of highway?

A. Quantity of conveyances

B. Assessment of traffic, financing, and environment

C. Cost of transportation and street parking

D. Noise generation and air pollution

3. Why do highways deteriorate before they come to the end of their service life?

A. Because the drainage structure of highways does not work.

B. Because the relevant agencies do not apply preventive maintenance treatments.

C. Because highways suffer from accumulated damage and adverse environmental effect.

D. Because residents attach great importance to environmental impact of constructing highways.

4. Which statement is true according to the text?

A. The first step of highway location is to draw longitudinal sections along possible alignment.

B. Duration of the storm is analyzed and calculated in the process of highway surfacing.

C. Old road surfaces, buildings and topsoil are compacted in the process of highway construction.

D. In the past, cobblestones and gravel highway surfaces are extensively used.

5. What is the main idea of the text?

A. Romans constructed roads because of military campaigns.

B. The drainage structure of highways plays an important role in highway engineering.

C. Highway construction processes include excavation, filling, compacting, and installation.

D. It summarizes the history of highway construction and gives a detailed explanation of highway engineering.

Exercise 2

Fill in the blanks with the words given below. Change the form where necessary.

| maintenance | adjacent | pavement | drainage | asphalt |
| flexible | removal | excavation | eliminate | prolong |

1. Our farm land is _____ to the river.
2. The news _____ doubts about the company's future.
3. There were three _____ in the first round of the competition.
4. The roads become really terrible, some of they are laid with _____ and gravel.
5. The operation could _____ his life by two or three years.
6. The tyres of Lenny's bike hissed over the wet _____.
7. Computers offer a much greater degree of _____ in the way work is organized.
8. The body was discovered when builders _____ the area.
9. These cuts have left the _____ system vulnerable when faced with heavy rains.
10. The school pays for heating and the _____ of the building.

Exercise 3

Translate the following sentences into Chinese.

1. The construction of modern highways in China contributes to the rapid development of national economy and improvement of people's living standard.

2. In today's culture of minimizing the highway construction impact on environment and society as a whole, a great deal of extra effort and time is expended to develop highway design.

3. Improved design methodologies, enhanced analysis techniques and the use of computer-assisted drafting have significantly improved the overall quality of highway construction.

_____

_____

4. After traffic, financing, and environmental considerations have been addressed, alignment of the highway is set out by a surveyor.

_____

_____

5. The economic feasibility of many types of road surfaces depends heavily on the cost and availability of building materials.

_____

_____

课文译文

公路工程

路指的是连接两个地点的大道，它能够允许人或某种交通工具（如马、马车、机动车等）通行。道路修建的起源可以追溯到罗马时代。出于军事原因，罗马道路以其直线性和耐久性为特点。通常情况下，罗马道路的表面高度要比实际地面高出一米甚至更多，这样，罗马人可以清晰地观察邻近地区的情况。这样一来，"公路"一词便出现了。

通常情况下，道路可分为公路、城市道路、厂矿道路、林区道路和乡村道路五种类型。公路主要被用于城市间的汽车运输。技术的进步和各国经济交流活动的增加促进了现代公路的发展。高质量的公路系统能通过减少行车时间和成本的方式进一步促进国家经济的发展和人民生活水平的提高。公路工程是土木工程学的一个分支，主要包括公路规划、公路设计、公路施工和公路维护四个部分。

在设计和修建一条新的公路前，必须对交通、财政和环境做整体规划和评估。公路工程师采集一系列交通信息（如交通流量、分布、特征等），并对某地区在一段时间内

（通常为 20 年）的交通需求进行评估。通过分析现有交通情况的数据及其可预知的变化，公路工程师才能决定何种公路建设才能满足当地的交通需求。

资金是否充足决定了工程是一次性完成还是分阶段施工。国家通过立法吸引个人和企业投资，促进公路建设的发展。工程师通过分析公路的经济可行性选取最佳规划方案。通常，公路使用者和所有权毗邻的所有者能获得一些好处，如降低运输费用、改善停车环境和提升所有权价值。

如今，居民非常重视公路建设对环境产生的影响。不合理规划或修建的公路产生的不利环境影响主要包括噪声、空气污染、水污染和自然景观的破坏。因为环境问题，不少工程被迫延期完工甚至被取缔。因此，公路工程师需通过采用环境影响评价方法充分了解工程施工对自然环境可能产生的影响并作出相应评估。

公路工程项目的详细设计主要涉及定线、路面设计和排水设计。当公路工程项目规划完成后，工程师开始着手对公路进行定线。首先，工程师对相关地区进行全面观察。然后，工程师沿着可能的线向绘制纵断面，并选择可能的路线。当然，所有的路线必须符合工程设计标准的要求。随后，定位器被用来对现场进行实地勘察，以寻找可行的路线并定位施工中可能遇到的障碍。通过仔细分析和勘察每一个可能路线，最终选定最佳路线。

路面指的是其顶部铺以耐用的筑路材料，并能承载客流量的道路结构层。以前鹅卵石路面和碎石路面随处可见，现今，它们基本上被沥青路面和混凝土路面取代了。路面结构层通常由面层、基层和垫层三部分组成。所有的坚硬路面都可以划分为柔性路面和刚性路面两类。柔性路面的表层都是沥青类材料，由于在交通荷载的情况下，这种路面结构会发生弯曲，所以，这种路面叫作"柔性路面"。刚性路面的表层是水泥混凝土。因为水泥混凝土刚度较大，所以刚性路面比柔性路面更加坚硬。路面的性能取决于各组成部分的性质，这就要求公路工程师对其做出准确的评估。

排水指的是将地表水和地下水排除。地表排水是指通过修建一条有路拱的路，使路肩及其附属区域倾斜，将水导向天然排水系统或导向有集水箱和地下管道的暴雨排水系统。如果使用了暴雨排水系统，公路工程师必须考虑预期的最大排水率、暴雨持续时间和每个集水箱的容量。当设计地下排水设施时，工程师必须考虑排水区域可能出现的最大降雨量和最大流速，然后推算出所需排水结构的负荷量。

公路施工主要包括挖掘、填筑、压实和安装四个步骤。首先，旧的路面、房屋和表层土被清除，填筑采用分层填筑压实方法。根据设计标准要求，在填筑过程中，每一层填料被摊铺并压实。气胎压路机压实公路下方的土壤以防沉降。公路施工的最后步骤是安装防撞栏、交通标志及其他路面标记。

公路维护包括路面、排水设施和交通控制设施的维修和保养。公路性能衰减的主要原因是车辆造成的累积损伤和环境的影响，如冻胀和热裂解。公路有明确的服务年限。几乎所有的公路在其使用寿命结束之前都需要某些种形式的养护。相关部门不断采取预防性养护措施来延长公路寿命。

References

[1] 段兵延. 土木工程专业英语[M]. 武汉：武汉理工大学出版社，2003.

[2] 董祥. 土木工程专业英语[M]. 南京：东南大学出版社，2011.

[3] Ricketts J T, Loftin M K, Merritt F S. Standard Handbook for Civil Engineers. McGraw-Hill Education, 2004.

Unit 7 Bridge Construction

Lead in

Identify the pictures below. Match them with the words in the box.

| suspension | arch | rigid beam |

| | | |

Section A

Dialogue 1

Several equipment operators are needed for the Bridge project. After the job vacancies have been advertised in the newspaper, the job seekers come to the site office for interviews. Mr. Zhang is receiving them.

J = Job Seeker   Z = Zhang

J: Good morning, I am asking about your advertisement for operators.

Z: What kind of job are you looking for?

J: I want to work here as a **bulldozer** operator.

Z: I am sorry, the vacancies have been filled. You should have tried earlier.

J: What a pity!

Z: If you like, you can fill in this application form and leave us your telephone number. As soon as we have a job that suits you, we will give you a call.

J: OK. Please call me once you have a job for me. I am quite skillful and I really want to work here.

Z: Right. I hope we'll soon have a job for you.

J: Thanks a lot.

New Words

vacancy	['veɪkənsi] n.	being unoccupied; an empty area or space 空缺；空位
bulldozer	['bʊldəʊzə(r)] n.	large powerful tractor; a large blade in front flattens areas of ground 推土机

Phrases and Expressions

equipment operator 设备操作工

application form 申请表

Dialogue 2

The phone rings.

Se = Secretary   S = Saito   H = Huang

Se: COCC(China Overseas Construction Company) office. Can I help you?

S: Hello, may I speak to the person in charge of equipment?

Se: Just a minute, please.

(calls Mr. Huang) Mr. Huang, you're wanted on the phone.

H: Hello, this is Huang.

S: Hello, Mr. Huang. This is a long distance call from USA Branch Office of Furukawa. My name is Saito. Our **headquarters** in Japan have asked me to contact you about the training of your operators for the rock drill jumbos you've bought from us. When do you want us to send our engineer to you?

H: We plan to do the training from 15th to 27th of this month, so I want that your engineer can arrive here on 13th or 14th.

S: No problem.

H: Does the engineer speak English? No one here can speak Japanese.

S: We will send Mr. Sunaga to you. He's been staying in the United States for nearly 3 years and speaks English fluently. Don't worry about it. Could you go to the airport to pick up Mr. Sunaga?

H: Sure.

S: OK, Mr. Huang, if you need any help, just let me know.

H: Thank you Mr. Saito.

New Words

headquarters [ˌhedˈkwɔːtəz] n. (usually plural) the office that serves as the administrative center of an enterprise 总部

Phrases and Expressions

branch office 分公司
drill jumbo 凿岩钻车，台车

Exercise 1

Decide whether the following statements are true (T) or false (F) according to the dialogues.

☐ 1. The job seeker is an equipment operator.

☐ 2. There are still vacancies for bulldozer operator.

☐ 3. COCC has bought rock drill jumbos from Saito's company.

☐ 4. Saito is a staff of the headquarters of Furukawa.

☐ 5. Mr. Sunaga will be picked up by Mr. Huang.

Exercise 2

Oral practice.

Directions: Pair work. Use the questions below to interview your partner and then change roles.

1. How long have you been working as an equipment operator?

_____

_____

2. Why did you quit your previous job?

_____

_____

3. Would you please introduce the steps before starting the engine?

_____

_____

4. When will you be available for work?

_____

_____

5. How much do you want for this position?

_____

_____

Exercise 3

Practical Activity.

Directions: Work in pairs. You are Mr. Zhang, and your company is recruiting dump truck drivers. Make an interview with your partner. The result of the discussion will be reported to the whole class.

Section B

Bridge

Bridge is a structure that **spans** obstacles, such as rivers and valleys, to provide a roadway for traffic. By far the majority of bridges are designed to carry automobile or railroad traffic, but some are intended for **pedestrians** only. A number of **aqueduct** bridges, mostly erected in Europe in the 19th century, carry canals and their **barge** traffic; and at least one bridge, at New York City's Kennedy Airport, serves to carry taxiing aircraft over a highway.

The first bridges built by man probably resembled those still being constructed by primitive peoples in isolated regions. The tools and building skills of early man, like those of primitive peoples today, were so elementary that he was undoubtedly forced to use easily transportable materials that could be put in place with a minimum of forming and shaping.

The first innovation beyond the primitive bridge forms is believed to have occurred in ancient China and then spread to India. To bridge streams wider than a single tree length, the Chinese and Indians used two piles of tree trunks, building toward the stream center from each bank. In each arm of the structure, the logs were piled on top of one another with a slight upward **slant**, with each layer projected several feet beyond the one immediately below it. For stability, each pile of timbers was **anchored** by a massive pile of stone on each bank. Near midstream, the gap between the ends of the arms was closed by the addition of a simple beam between the two ends. In this kind of structure, a crude **cantilever** bridge, the achievable span length is made greater by the addition of a central section between the free ends of arms.

As early as 4000 B.C. in Mesopotamia and 3000 B.C. in Egypt, **overlapping** horizontal layers of stone or sun-baked brick were used instead of overlapping timbers. This construction, which looks like a masonry arch that is stepped rather than smooth on the underside is called a corbel arch. To change the corbel arch into a true arch, it was only necessary to **reorient** the inner stones so that they formed a smooth curve. The true arch, which was used as early as 3500 B.C., is much stronger than the corbel arch.

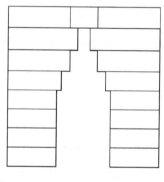

corbel arch

The true masonry arch was efficient, and durable. It could span small rivers by multiple arches resting on **piers**; moreover, it was generally gracious in appearance. These qualities were hard to match by any previous form of construction. The true masonry arch was widely used in bridge construction by both the ancient Chinese and the Romans; it remained in wide use until the 19th century.

Basically there are just four types of structures that can be used to bridge a stream or other obstacle: rigid beams, cantilevers, arches, and suspension systems.

Rigid Beam　　The simplest way—probably the first used—is to lay a rigid beam across the stream so that its two ends rest on opposite banks. The rigid beam in this type of bridge may be a shaped wooden beam, a **girder** of steel or **concrete**, or even a complex **truss**. The span of a rigid beam type of bridge can be increased by building intermediate piers and bridging the gaps between the piers with several beams. The materials used for rigid beams must be able to withstand both compression and tension. The double requirement stems from the fact that when a load is placed on a rigid beam, the beam bends—in spite of its name. As a result of bending, the upper part of the beam is compressed and the lower half is stretched. If its compressive strength is too low, it will **buckle**; if its **tensile** strength is too low, it will break.

Cantilever Often it is not feasible to construct long span bridges by means of intermediate piers. In deep swift-running rivers or in soft subsoils for example, it may be difficult to construct the piers or to make them deep enough to reach a firm foundation rock. In such cases the span of a rigid-beam structure can be extended simply by using two beams —one extending from each bank—with one end of each beam firmly anchored to its foundation, rather than simply resting on it as in ordinary rigid-beam construction. Each of the anchored beams in this kind of structure is called a cantilever. Perhaps the simplest familiar example of a single cantilever is the familiar diving board. In an ordinary cantilever bridge, the gap between the ends of the cantilevers is closed, providing a continuous deck for the roadway, but if the bridge were cut in two at the point of closure, each cantilever would support itself. Usually the gap between the cantilevers is closed by means of a rigid beam, thus extending the span of the cantilevers.

Suspension Suspension bridges can span even greater distances without intermediate piers than cantilever bridges. The supporting members of a suspension bridge are continuous flexible cables, with each cable anchored at both of its ends. The simplest example of the suspension bridge is the **circus aerialist**'s tightrope, and primitive suspension bridges were often no more than several such tightropes tied together to provide both hand and foot holds. In modern suspension bridges, a level roadway is provided by stringing the cables overhead on high towers and suspending the separate roadway below.

Arch The arch is in a sense the opposite of a suspension cable. Where the suspension cable hangs freely from its supporting towers, the arch curves rigidly upward from its **abutments**. Because of this difference in shape, the suspension cable tends to pull its **anchorages** together, while the arch tends to push its abutments apart. For this reason suspension cables must be able to withstand stretching, while arches must be made of materials that can withstand compression. Because tensile strength is not necessarily required for arch construction, arch bridges can be made of bricks or stone blocks that are held together by the compressive force of the arch. Such materials are useless for the other basic bridge structures.

In an arch bridge, the loads are transmitted vertically down from the roadway until they intersect the arch form. At the arch boundary, the loads are redirected along the path of the arch, which is in a state of simple compression. The compressive thrust forces are then transmitted to the ground through the abutments or piers. The arch, with its simple and elegant structures, has become a classic bridge **configuration**.

New Words

span	[spæn] v.	to cover or extend over an area or time period 跨越
pedestrian	[pəˈdestrɪən] n.	a person who travels by foot 行人；步行者
aqueduct	[ˈækwɪdʌkt] n.	a conduit that resembles a bridge but carries water over a valley 渡槽；导水管；沟渠
barge	[bɑːdʒ] n.	a flatbottom boat for carrying heavy loads (especially on canals) 驳船；游艇
slant	[slænt] n.	degree of deviation from a horizontal plane 倾斜
anchor	[ˈæŋkə(r)] n.&v.	a mechanical device that prevents a vessel from moving 锚；抛锚停泊 secure a vessel with an anchor 抛锚；使固定
cantilever	[ˈkæntɪliːvə(r)] n.	projecting horizontal beam fixed at one end only 悬臂
overlap	[ˌəʊvəˈlæp] v.	coincide partially or wholly; extend over and cover a part of 部分的同时发生；部分重叠
reorient	[ˌriːˈɔːrɪent] v.	set or arrange in a new or different determinate position 使适应；再调整
pier	[pɪə(r)] n.	a support for two adjacent bridge spans 桥墩

girder	[ˈgɜːdə(r)] n.	a beam made usually of steel; a main support in a structure 大梁，纵梁
concrete	[ˈkɒŋkriːt] n.	a strong hard building material composed of sand and gravel and cement and water 混凝土
truss	[trʌs] n.	a framework of beams (rafters, posts, struts) forming a rigid structure that supports a roof or bridge or other structure 桁架
buckle	[ˈbʌkl] v.	bend out of shape, as under pressure or from heat 变弯曲
tensile	[ˈtensaɪl] adj.	capable of being shaped or bent or drawn out 拉力的；可伸长的；可拉长的
circus	[ˈsɜːkəs] n.	a travelling company of entertainers, including trained animals 马戏；马戏团
aerialist	[ˈeərɪəlɪst] n.	an acrobat who performs in the air (as on a rope or trapeze) 高空杂技演员
abutment	[əˈbʌtmənt] n.	a masonry support that touches and directly receives thrust or pressure of an arch or bridge 桥礅，桥台
anchorage	[ˈæŋkərɪdʒ] n.	place for vessels to anchor; the act of anchoring 锚地；下锚
configuration	[kənfɪɡəˈreɪʃn] n.	an arrangement of parts or elements 结构；外形

Phrases and Expressions

be intended for 打算为……所用；预定给

put in place 到位；落实到位；正在实施

masonry arch 砌石拱

corbel arch 突拱

be hard to match 无可匹敌，无法与之媲美，难以企及

rigid beams 刚架桥

no more than 只是；仅仅

in a sense 从某种意义上说

Exercise 1

Choose the best answer to each of the following questions.

1. Which of the following is introduced in the text?

   A. Rigid beams

   B. Cantilevers

   C. Arches

   D. All of the above.

2. The beam in rigid beams could be the following materials **EXCEPT** _____.

   A. girder of steel

   B. stone

   C. concrete

   D. complex truss

3. What does the passage say about Cantilevers?

   A. It is possible to construct long span bridges by means of intermediate piers.

   B. The diving board is a kind of cantilever.

   C. It is no longer difficult to construct the piers in soft subsoils now.

   D. The gap between the cantilevers is closed by concrete.

4. Which of the following is **NOT** true?

   A. Suspension bridges are supported by continuous flexible cables.

   B. Primitive suspension bridges were only several tightropes tied together.

   C. Cantilever bridges can span greater distances without intermediate piers than suspension bridges.

   D. Each cable of suspension bridges should be anchored at both of its ends.

5. How do the loads be transmitted in an arch bridge?

   A. At the arch boundary, the loads are transmitted along the path of the arch.

   B. The loads are transmitted horizontally from the roadway.

   C. The compressive thrust forces are transmitted to the river through piers.

   D. The loads are transmitted vertically up from the abutments.

108 土木工程实用英语 Practical English for Civil Engineering

Exercise 2

Fill in the blanks with the words given below. Change the form where necessary.

| continue | close | anchor | use | direct |
| span | basic | bridge | suspend | tensile |

Bridge is a structure that _____ obstacles, such as rivers and valleys, to provide a roadway for traffic. _____ there are just four types of structures that can be used to bridge a stream or other obstacle: rigid beams, cantilevers, arches, and suspension systems. The span of a rigid beam type of bridge can be increased by building intermediate piers and _____ the gaps between the piers with several beams. The materials _____ for rigid beams must be able to withstand both compression and _____. In an ordinary cantilever bridge, the gap between the ends of the cantilevers is closed, providing a _____ deck for the roadway, but if the bridge were cut in two at the point of _____, each cantilever would support itself. The supporting members of a suspension bridge are continuous flexible cables, with each cable _____ at both of its ends. In modern suspension bridges, a level roadway is provided by stringing the cables overhead on high towers and _____ the separate roadway below. In an arch bridge, the loads are transmitted vertically down from the roadway until they intersect the arch form. At the arch boundary, the loads are _____ along the path of the arch, which is in a state of simple compression. The compressive thrust forces are then transmitted to the ground through the abutments or piers.

Exercise 3

Translate the following sentences into Chinese.

1. By far the majority of bridges are designed to carry automobile or railroad traffic, but some are intended for pedestrians only.

_____

_____

2. To bridge streams wider than a single tree length, the Chinese and Indians used two piles of tree trunks, building toward the stream center from each bank.
_____
_____

3. This construction, which looks like a masonry arch that is stepped rather than smooth on the underside is called a corbel arch.
_____
_____

4. It could span small rivers by multiple arches resting on piers; moreover, it was generally gracious in appearance. These qualities were hard to match by any previous form of construction.
_____
_____

5. The simplest example of the suspension bridge is the circus aerialist's tightrope, and primitive suspension bridges were often no more than several such tightropes tied together to provide both hand and foot holds.
_____
_____

课文译文

桥 梁

桥梁是一种跨越河流和山谷等障碍的结构形式，提供便利的交通。到目前为止，大部分桥梁都是公路桥或铁路桥，也有少部分人行天桥。大量的水道桥于19世纪在欧洲建成，用于通水和通航；用于把滑行飞机拖到跑道上的桥少则有一座，位于纽约肯尼迪机场。

人类建成的第一座桥大概跟那些与世隔绝的原始民族仍然在建的桥类似。早期，人类使用的工具和建筑技术，同今天的原始民族一样，非常初级。所以，早期人类不得不使用易于运输的材料，只需最少的加工成型，即可安装使用。

原始桥梁的首次革新被认为出现在古代中国，随后传入印度。在河面宽度超过单根树木长度的地方，中国人和印度人用两堆树干从两岸向河心而建。桥的两臂，原木层层堆叠，并微微向上倾斜，每一层都比下面一层高几英尺。为了增加稳定性，每堆木材都靠两岸一堆大而重的石头锚固，接近河心的位置，桥梁两臂之间用简支梁连接。这种结构，在两臂自由端间加横梁可让一座粗糙建成的悬臂桥达到更大的跨度。

早在公元前 4000 年的美索不达米亚和公元前 3000 年的埃及，水平堆叠的石头或晒干的砖代替了层叠的木材，这种结构叫作突拱，看起来是阶梯式而不是拱面平滑的砌石拱。若要让突拱（也称"假拱"）变为"真拱"，只需调整内部的石头，使之形成平滑曲面。真拱比突拱更坚固，早在公元前 3500 年就已开始使用。

真拱石拱桥建成速度快，经久耐用。通过桥墩连拱便可跨越小河，且外观优雅别致，质量过硬，先前的任何结构形式均无可匹敌。在古代中国和古罗马，石拱桥被广泛地用于桥梁建设，直到 19 世纪仍在广泛使用。

大致来说，可用作水面上或其他障碍物上的桥有四类：刚架桥、悬臂桥、拱桥和悬索桥。

刚架桥　最简单也可能是最早使用的桥即刚架桥，将刚性梁横跨河面，两端固定在河两岸。这种桥的刚性梁可以是加工过的木梁、钢筋混凝土梁，或更复杂的桁架。河中建桥墩，用横梁将桥墩连接起来，可增加刚架桥的跨径。刚架桥的材料必须能够承受住压力和拉力，这两方面的要求是因为尽管名字叫刚架桥，但桥上有荷载时，横梁会弯曲。在弯曲的作用下，横梁上半部受压，下半部被拉伸，抗压强度太低就会塌陷，抗拉强度太低则会断裂。

悬臂桥　中间建桥墩的方式对大跨度桥通常不可行。比如，在水深流速急的河流或软泥中，很难建桥墩或使桥墩深达到基岩层。这种情况，刚架桥结构的跨径用两根横梁就可以延伸——从两岸各伸出一根梁，对两根梁的端部基底进行锚固，而不是像普通刚架桥那样简单固定，每根锚固的梁就叫悬臂，最简单熟悉的单悬臂可能就是跳水板了。在普通的悬臂式桥梁中，悬臂端之间的间距是闭合的，提供了连续的桥面道路。假如把桥在闭合点断开，每根悬臂也都能自支撑。悬臂间距通常用刚性梁闭合，这样便延伸了悬臂桥的跨度。

悬索桥　在没有中间桥墩的情况下，悬索桥比悬臂桥能跨越更大的距离。悬索桥的支撑体系是连续柔韧的缆绳，每根缆绳两端进行锚固。悬索桥最简单的例子是杂技演员高空走钢丝用的钢丝，原始的悬索桥不过是很少的几根这种钢丝系在一起，供手扶和站立。现代悬索桥体系中，水平道路由上部索塔悬挂的成排缆绳承重，路面独立悬于下方。

拱桥 从某种意义上说，拱桥与悬索缆绳相反，悬索缆绳从索塔向下自由悬挂，拱桥从两端桥台向上弯曲。由于桥的形状不同，悬索缆绳把桥墩往一处拉，而拱桥把桥台往两边压。因此，悬索桥的缆绳必须能抗拉，而拱桥的材料则须能抗压。材料的抗拉强度对拱形建筑不是硬性要求，所以拱桥可以用砖或石头建造，砖或石头通过拱传递的压力结合在一起。这些材料在其他几个基本桥梁结构中毫无用处。

在拱桥结构中，荷载从桥面垂直传递下来，直到与拱相交。在拱边缘处，荷载会改变力的传递路径，变为一种简单的受压状态。有压缩力的推力通过桥台或桥墩传到地面。拱桥，以其简单优美，成为了一种经典的桥梁结构。

References

[1] 张水波，刘英. 国际工程管理实用英语口语：承包工程在国外[M]. 北京：中国建筑工业出版社，1997.

[2] Corbel arch [EB/OL]. (2017-02-10) http://en.wikipedia.org/wiki/corbel_arch.

[3] 桥梁[EB/OL]. (2017-02-02) http://wenku.baidu.com/view/e7f83f4acbaedd3383c4bb4cf7ec4afe04a1b1f6.

Unit 8　Rail Engineering

Lead in

Identify the pictures below. Match them with the words in the box.

| HSR | rolling stock | maglev |

Section A

Dialogue 1

Mr. Chen, officer of State Railways Administration, has made an appointment with Mr. Liao, project manager of China Railway Engineering Corporation, to visit the site today. Mr. Chen arrives at the site office where Mr. Liao and Mr. Zhang are expecting him.

L= Mr. Liao　　C=Mr. Chen　　Z=Mr. Zhang

L: Welcome to our project, Mr. Chen.

C: Thank you.

L: This is my assistant, Mr. Zhang.

C: Glad to see you, Mr. Zhang.

Z: Glad to meet you too, Mr. Chen. We are honored by your visit. Shall we go to the site now?

C: OK.

Z: I am afraid the road to the site is not so good. Let's take my Bronco; it's a four-wheel drive, very powerful. I hope you don't mind.

C: Of course not, if we are going to the job site. How far is it from here?

Z: About two and a half miles.

On the way to the site…

C: (See workers laying **tracks** alongside the road) I suppose they are building the new railway.

Z: Actually this is a part of the whole project. We have **sublet** it to a local **contractor**. Besides this new railway project, we are constructing a new station.

C: That's very impressive. What are those two big things?

L: They are **cement silos,** where **bulk** cement is stored.

C: When will you complete the whole project?

L: It should be completed by December next year according to the contract.

Z: I think we will complete the project two months ahead of time if everything goes smoothly.

C: I am glad to hear that.

New Words

track	[træk] n.	railway tracks are the rails that a train travels along 轨道
sublet	[ˌsʌbˈlet] vt.	lease or rent all or part of (a leased or rented property) to another person 转租；转包给
contractor	[kənˈtræktə(r)] n.	a person or company that does work for other people or organizations 承包人；订约人

cement	[sɪˈment] n.	a grey powder which is mixed with sand and water in order to make concrete 水泥
silo	[ˈsaɪləʊ] n.	a cylindrical tower used for storing silage 筒仓
bulk	[bʌlk] adj.	in large quantity 散装的；大批的，大量的

Phrases and Expressions

Bronco "野马"牌汽车
four-wheel drive 四驱（汽车）

Dialogue 2

Jane, a journalist, spoke to Dr. Ruth, professor of Logistics and Transport of university, and also an adviser to Thailand's transport department.

J= Jane   R= Dr. Ruth

J: Explain why trains and tracks in China are currently not **compatible** with those of Southeast Asia.

R: There is no standard width for railways around the world. For China, it's 1.44 meters, it's wider, so **theoretically** they can have higher speed trains and can carry more freight. In many Southeast Asia countries, they have the one-meter **gauge**. It's an old **relic** from colonial times and it's a system that's promoted by ASEAN.

J: So how would that work with this new rail system that China wants to **implement**?

R: So what it means is that the country like Thailand has two rail systems, a one-meter gauge or system which is very important to connect the neighboring countries and another specific system just to connect cities within China.

J: What kind of benefits will China and Thailand have to build this new rail line?

R: From China's perspective, getting access into **Laos** is very important because they have a number of mining **concessions**, and linking with Thailand gives them another access to the sea. For Thailand, it is a good opportunity to stimulate its economy.

New Words

compatible	[kəmˈpætəbl] adj.	things work well together or can exist together successfully 兼容的；和谐的，协调的

theoretically	[ˌθɪəˈretɪklɪ] adv.	Although something is supposed to be true or to happen in the way stated, it may not in fact be true or happen in that way. 理论上
gauge	[geɪdʒ] n.	a device that measures the amount or quantity of something and shows the amount measured. 评估；尺度，标准；测量仪器
relic	[ˈrelɪk] n.	a relic belonged to an earlier period but have survived into the present. 遗迹，遗物；废墟
implement	[ˈɪmplɪmənt] vt.	to ensure that what has been planned is done 实施，执行；使生效；落实（政策）
concession	[kənˈseʃn] n.	a special right or privilege that is given to someone （尤指由政府或雇主给予的）特许权；承认或允许

Phrases and Expressions

be compatible with... 与……相兼容；与……合适
ASEAN Association of Southeast Asian Nations 东南亚国家联盟
Laos 老挝

Exercise 1

Decide whether the following statements are true (T) or false (F) according to the dialogues.

☐ 1. The whole project refers to the construction of the railway and the new station.

☐ 2. The railway project will be completed by December next year if everything goes smoothly.

☐ 3. Trains and tracks in China are quite compatible with those in Thailand.

☐ 4. Theoretically China can have higher speed trains and can carry more freight.

☐ 5. If this new rail line is built, both China and Thailand will reap some benefits.

Exercise 2

Oral practice.

Directions: Pair work. Use the questions below to interview your partner and then change roles.

1. How far is it from the office to the project site?
_____
_____

2. When will you complete the whole project?
_____
_____

3. Explain why trains and tracks in China are currently not compatible with those in Southeast Asia?
_____
_____

4. What kind of benefits will China and Thailand have, if this new rail line is built?
_____
_____

5. What does "ASEAN" refer to?
_____
_____

Exercise 3

Practical Activity.

Directions: Work in pairs. What is the definition of "rolling stock" and how many categories are there? Please conduct a research on the Internet, and discuss with your classmates. Then the result of the discussion will be reported to the whole class.

Section B

High-speed Rail (HSR) in China

High-speed rail (HSR) in China is the longest HSR system in the world extending to 29 of the country's 33 **provincial**-level entities. The network consists of newly built passenger-dedicated lines (PDLs) and intercity lines along with upgraded mixed passenger and **freight** lines. The newly built PDLs without including intercity and upgraded passenger and freight lines currently account for 20,000 km of service routes, a length that is more than the rest of the world's high-speed rail tracks combined. The addition of PDLs and other high-speed lines is ongoing with the network of PDLs alone set to reach 38,000 km in 2025. High-speed rail service in China was introduced on April 18, 2007 and has become **immensely** popular with an annual **ridership** of over 1.44 billion in 2016, making the Chinese HSR network the most heavily used in the world. Notable lines include the world's longest line, the 2,298 km Beijing – Guangzhou High-Speed Railway and the Shanghai Maglev, the world's first high-speed commercial magnetic **levitation** (maglev) line and the only non-**conventional** track line of the network.

Nearly all high-speed rail lines and rolling stock are owned and operated by the China Railway Corporation, the state enterprise formerly known as the Railway Ministry, which has **overseen** the HSR building boom with generous funding from the Chinese government's economic **stimulus** program. The pace of high-speed rail **expansion** slowed for a period in 2011 after a fatal high-speed railway accident near Wenzhou, but has since **rebounded**. Though the system as a whole is considered successful, concerns about HSR safety, high-ticket prices, low ridership, financial sustainability and environmental impact of high-speed rail continue to exist for several projects.

China's early high-speed trains were imported or built under technology transfer agreements with foreign train-makers including Alstom, Siemens, Bombardier and Kawasaki Heavy Industries. Chinese engineers then re-designed internal train components and built **indigenous** trains. Today, most of China's high-speed rail trains are thus made in China by the CRRC.

History

The earliest example of higher-speed commercial train service in China was the Asia Express, a luxury passenger train that operated in Japanese-controlled Manchuria from 1934 to 1943. The steam-powered train, which ran on the Southern Manchuria Railway from Dalian to Xinjing (Changchun), had a top commercial speed of 110 km/h and test speed of 130 km/h. It was faster than the fastest trains in Japan at the time. After the founding of the People's Republic of China in 1949, this train model was renamed the SL-7 and was used by the Chinese Minister of Railways.

State planning for China's current high-speed railway network began in the early 1990s. In December 1990, the Ministry of Railways (MOR) **submitted** a proposal to build a high-speed railway between Beijing and Shanghai to the National People's Congress. At the time, the Beijing Shanghai Railway was already at capacity, and the proposal was jointly studied by the Science & Technology Commission, State Planning Commission, State Economic & Trade Commission, and the MOR. In December 1994, the State Council **commissioned** a feasibility study for the line. Policy planners debated the necessity and economic viability of high-speed rail service. Supporters argued that high-speed rail would boost future economic growth. Opponents noted that high-speed rail in other countries were expensive and mostly unprofitable. Overcrowding on existing rail lines, they said, could be solved by expanding capacity through higher speed and frequency of service. In 1995, Premier Li Peng announced that preparatory work on the Beijing Shanghai HSR would begin in the 9th Five Year Plan (1996-2000), but construction was not scheduled until the first decade of the 21st century.

The "Speed Up" Campaigns

In 1993, commercial train service in China averaged only 48 km/h and was steadily losing market share to airline and highway travel on the country's expanding network of expressways. The MOR focused modernization efforts on increasing the service speed and capacity on existing lines through double-tracking, **electrification**, improvements in grade (through tunnels and bridges), reductions in turn **curvature**, and installation of continuous **welded** rail. Through five rounds of "Speed-Up" campaigns in April 1997, October 1998, October 2000, November 2001, and April 2004, passenger service on 7,700 km of existing tracks was upgraded to reach sub-high speed of 160 km/h.

A notable example is the Guangzhou-Shenzhen Railway, which in December 1994 became the first line in China to offer sub-high-speed service of 160 km/h using domestically produced DF-class diesel **locomotives**. The line was electrified in 1998, and Swedish-made X2000 trains increased service speed to 200 km/h. After the completion of a third track in 2000 and a fourth in 2007, the line became the first in China to run high-speed passenger and freight service on separate tracks. In 2007, the Guangshen Railway became the first in the country to have four tracks, allowing faster passenger train traffic (on dedicated tracks third and fourth from the right) to be separated from slower freight traffic (on tracks second and fifth from the right).

The completion of the sixth round of the "Speed-Up" Campaign in April 2007 brought HSR service to more existing lines: 423 km capable of 250 km/h train service and 3,002 km capable of 200 km/h. In all, travel speed increased on 22,000 km, or one-fifth, of the national rail network, and the average speed of passenger trains improved to 70 km/h. The introduction of more non-stop service between large cities also helped to reduce travel time. The non-stop express train from Beijing to Fuzhou shortened travel time from 33.5 to less than 20 hours. In addition to track and scheduling improvements, the MOR also deployed faster CRH series trains. During the Sixth Railway Speed Up Campaign, 52 CRH trainsets (CRH_1, CRH_2 and CRH_5) entered into operation. The new trains reduced travel time between Beijing and Shanghai by two hours to just less than 10 hours.

New Words

provincial	[prə'vɪnʃl] adj.	connected with the parts of a country away from the capital city 省份的；地方的
freight	[freɪt] n.	the movement of goods by trains, ships, or aeroplanes 货运
immensely	[ɪ'mensli] adv.	to emphasize the degree or extent of a quality, feeling, or process 极大地，广大地；无限地，庞大地
ridership	['raɪdə,ʃɪp] n.	the volume of passengers 乘客流（通）量
levitation	[,levɪ'teɪʃn] n.	the act of raising (a body) from the ground by presumably spiritualistic means 漂浮；浮起
conventional	[kən'venʃənl] adj.	following accepted customs and proprieties 传统的；依照惯例的
oversee	[,əʊvə'siː] vt.	in authority oversees a job or an activity, make sure that it is done properly. 监督，监视
stimulus	['stɪmjələs] n.	something that encourages activity in people or things 促进因素；刺激物
expansion	[ɪk'spænʃn] n.	is the process of becoming greater in size, number, or amount 扩张
rebound	[rɪ'baʊnd] vi.	return to a former condition 弹回；从诸如衰败或失望中恢复过来
indigenous	[ɪn'dɪdʒənəs] adj.	originating where it is found 本地的；本土的
submit	[səb'mɪt] vt.	send a proposal, report, or request to someone, and they can consider it or decide about it 提交，呈送
commission	[kə'mɪʃn] vt.	formally arrange for someone to do a piece of work for you 委任；授予
electrification	[ɪ,lektrɪfɪ'keɪʃn] n.	a house, town, or area is the connecting of that place with a supply of electricity 电气化
weld	[weld] vt.& vi.	join together by heating 焊接；使紧密结合
locomotive	[,ləʊkə'məʊtɪv] n.	a wheeled vehicle consisting of a self-propelled engine that is used to draw trains along railway tracks 机车；火车头

Unit 8　Rail Engineering　121

Exercise 1

Choose the best answer to each of the following questions.

1. Why did the pace of high-speed rail expansion slow down for a period in 2011?

    A. Because the corruption scandal of Chinese Railways Minister and a fatal high-speed railway accident happened.

    B. Because it has since rebounded.

    C. Because it concerns problems like HSR safety, high-ticket prices, low ridership, financial sustainability and environmental impact of high-speed rail.

    D. All of above.

2. The Asia Express was operated _____.

    A. in Japan

    B. in Southern Manchuria

    C. in Dalian

    D. in Xinjing

3. The construction of Beijing-Shanghai HSR took _____.

    A. 4 years

    B. 10 years

    C. 20 years

    D. not mentioned

4. The underlined word "deployed" in the last paragraph can be best replaced by _____.

    A. employed

    B. arranged

    C. disposed

    D. used

5. What is the main topic of this article?

    A. The history of high speed railway in China.

    B. A general introduction of high speed railway in China.

    C. The function of MOR.

    D. The feasibility of high speed railway project.

Exercise 2

Fill in the blanks with the words given below. Change the form where necessary.

| note | rebound | commerce | operate | suggest |
| round | promote | intercity | consist | upgrade |

High-speed rail (HSR) in China, _____ of newly built passenger-dedicated lines (PDLs) and _____ lines along with _____ mixed passenger and freight lines, is the longest HSR system in the world. _____ lines include the world's longest line, the 2,298 km Beijing – Guangzhou High-Speed Railway and the Shanghai Maglev, the world's first high-speed commercial magnetic levitation (maglev) line and the only non-conventional track line of the network. But, the pace of high-speed rail expansion slowed in 2011 and then it has _____.

The Asia Express, the first higher-speed _____ train service in China, was _____ in Japanese-controlled Manchuria from 1934 to 1943. It was faster than the fastest trains in Japan at the time. In 1990, the Ministry of Railways (MOR) _____ to build a high-speed railway between Beijing and Shanghai. At the time, the Beijing Shanghai Railway was already at capacity. Through five _____ of "Speed-Up" campaigns in April 1997, October 1998, October 2000, November 2001, and April 2004, passenger service on 7,700 km of existing tracks was _____ to reach sub-high speeds of 160 km/h. During the Sixth Railway Speed Up Campaign, 52 CRH trainsets (CRH1, CRH2 and CRH5) entered into operation. The new trains reduced travel time between Beijing and Shanghai by two hours to just less than 10 hours.

Exercise 3

Translate the following sentences into Chinese.

1. The network consists of newly built passenger-dedicated lines (PDLs) and intercity lines along with upgraded mixed passenger and freight lines.

_____

_____

2. High-speed rail service in China was introduced on April 18, 2007 and has become immensely popular with an annual ridership of over 1.44 billion in 2016, making the Chinese HSR network the most heavily used in the world.

_____

_____

3. The steam-powered train, which ran on the Southern Manchuria Railway from Dalian to Xinjing (Changchun), had a top commercial speed of 110 km/h and test speed of 130 km/h.

_____

_____

4. In 1993, commercial train service in China averaged only 48 km/h and was steadily losing market share to airline and highway travel on the country's expanding network of expressways.

_____

_____

5. The completion of the sixth round of the "Speed-Up" Campaign in April 2007 brought HSR service to more existing lines: 423 km capable of 250 km/h train service and 3,002 km capable of 200 km/h.

_____

_____

课文译文

中国高速铁路

中国拥有全世界最长的高铁线路，全国 33 个省级单位中有 29 个通了高铁。高铁线路包括新建的客运专线铁路（客运专线）和通过改建的、可同时容纳乘客和货物的城际线路。除开改建的城际客货运路线，新建的客运专线目前已经达到 20 000 千米的里程，长度比世界上剩下的高速铁路轨道加起来还多。新增的客运专线等高速铁路是单独客运专线网络的组成部分，预计单独的客运专线在 2025 年到达 38 000 千米。中国高铁自 2007 年 4 月 18 日正式投入运行以来受到了广泛的欢迎，2016 年高铁的年载客量超过了 14.4 亿人次，中国高铁也因此成为世界上最繁忙的路线。其中著名的路线有北京 - 广州高铁线和上海磁悬浮高速铁路。全长 2 298 千米的京广高铁是世界最长的高铁路线，上海的磁悬浮高铁是世界上首个投入商业运营的磁悬浮线路，也是整个高铁体系中唯一一个非传统轨道线路。

所有的高铁线路和运行车辆均由中国铁路总公司运营并所有，该公司属国有企业，其前身是铁道部。从中国政府的经济刺激计划中，铁道部运用政府大力的财政支持，负责推动高铁建设的热潮。2011 年，在惨烈的温州高铁事故后，中国高铁放慢了扩张速度，但此后又有所回升。尽管整体来讲高铁系统是成功的，但考虑到安全性，部分高铁项目仍存在高票价、低运量、财务可持续性低和高速铁路环境差等方面的问题。

中国早期的高速列车来自国外进口或与国外制造商签订技术转让协议建造，其中包括阿尔斯通、西门子、庞巴迪和川崎重工等列车制造商。后来中国工程师重新设计了列车内部部件，制造出了本土列车。因此今天大多数中国的高速铁路列车由 CRRC 在中国国内制造。

发展历史

中国高速列车投入商业运营的最早案例是亚洲快线，从 1934 年到 1943 年，这辆豪华列车奔驰在当时处于日本控制的满洲里。这辆蒸汽动力的火车从满洲里南部的大连驶往新京（长春），最高运营速度 110 千米/小时，最高测试速度 130 千米/小时。这速度比当时日本最快的火车还要快。1949 年在中华人民共和国成立后，这个列车型号改称为 SL-7，由中国铁道部部长所用。

建设中国目前的高速铁路网的计划开始于 20 世纪 90 年代初期。1990 年 12 月，铁道部提交了一份议案给全国人民代表大会，建议在北京和上海之间的修建高速铁路。那时，北京 - 上海铁路已经满负荷，此项提议由科学技术委员会、国家计划委员会、国家经济贸易委员会以及铁道部共同研究考察，1994 年 12 月，国务院委托对该线路实行可行性研究。政策制定者们争论着高铁服务的必要性和经济可行性。支持者认为高铁将促进未来经济增长。反对者指出在其他国家的高速铁路费用昂贵的，大多处于亏损。他们说道，现有铁路线路过度拥挤的问题可以通过提速和提高发车频率来解决。1995 年，李鹏总理宣布北京上海高铁将纳入第 9 个五年计划（1996 - 2000）中开展筹备工作，2000 年的头个十年开始施工建设。

"高铁提速"之路

1993 年，中国商业运营铁路运输平均时速为 48 千米，与航空运输和不断扩张的高速公路网络运输相比，渐渐丢失了市场份额。中国铁路局通过着眼于现代化铁路建设，提高电气化水平，提升铁路等级（通过隧道和桥梁），减少转弯半径，新建无缝铁路，增加现有双轨铁路线路的运输能力等方式，提高了运输速度。通过 1997 年 4 月，1998 年 10 月，2000 年 10 月，2001 年 11 月以及 2004 年 4 月的五轮"提速"，现有的 7 700 千米的客运线路改善升级，达到 160 千米每小时的准高速运行速度。

值得一提的是，1994 年 12 月，广深铁路作为中国第一条准高速铁路，采用了国产的 DF 级柴油火车头，时速可达 160 千米。1998 年，广深铁路线实现了电气化，采用瑞士制造的 X2000 列车，时速提高到 200 千米，此后，2000 年、2007 年广深铁路第三条和第四条铁轨相继完工，此线路成为了中国首条实行货物运输与高铁运输分离的线路。2007 年，广深高铁是我国国内第一家拥有四轨的线路，运送旅客的快车（右边第三和第四专用轨道）就与铁路货车（右边第二和第五轨道）实现了轨道分离。

2007 年 4 月，中国铁路第六轮"提速"圆满完成，多条线路实现了高铁运输，时速达 250 千米的线路累计 423 千米，时速可达 200 千米的线路 3 002 千米。运输速度增长的铁路总共 22 000 千米，即全国铁路运输网络的五分之一。全国客运铁路平均运输速度提高到 70 千米每小时。大型城市直达列车的推广也有利于缩短行驶时间。北京至福州直达快速列车将运行时长从 33.5 小时减少到少于 20 小时。除了在轨道和运行上的调整，国家铁路总局还引入了运行速度更快的 CRH 型列车。在第六轮的提速中，52CRH 列车组（CRH_1，CRH_2 和 CRH_5）投入运行，将北京至上海线路的运行时间减少了 2 小时，总运行时间控制在 10 小时以内。

References

[1] 井国庆. 铁道工程专业英语[M]. 北京：北京交通大学出版社，2013.

Keys to Exercises

Unit 1

Lead-in

casting yard; distillation tower; crane

Section A

Exercise One

1. F   2. T   3. F   4. T   5. F

Exercise Two

1. Certainly. This is Mr. Du, a Chinese builder.

2. At least five months.

3. Sure. Hello, fellow workers! This is our new fellow from China, Mr. Hu.

4. Mr. Wang is in charge of the installation work.

5. Yes, that hoist is made in Japan.

Exercise Three

Z=Mr. Zeng   H=Mr. Hu   G=Mr. Gao

Z: Good morning, Mr. Hu. Nice to meet you.

H: Good morning. Nice to meet you, too. So, you must be Mr. Zeng?

Z: Exactly. Welcome to the construction site in Kenya.

H: Thank you very much.

Z: I'll show you around the construction site. Follow me, please.

H: Thank you.

Z: Here we are.

H: Would you please introduce me to the section chief on the building site?

Z: Sure. This is Gao Hu, a builder from Si Chuan Province.

H: How do you do, Mr. Gao? Nice meeting you.

G: How do you do, Mr. Hu? Nice meeting you, too.

Z: Mr. Gao, this is Mr. Hu. He is an engineer from China State Construction

Engineering Corporation and he'll take charge of the construction project here.

G: Welcome to our building site, Mr. Hu. How long will you work here?

H: Until the construction project is completed. I hope we'll have a very smooth cooperation here.

G: I hope so.

Section B

Exercise One

1. D   2. A   3. C   4. B   5. C

Exercise Two

1. supervise

2. hazardous

3. oversee

4. well-being

5. respectively

6. contracts

7. combustible

8. eliminated

9. converting

10. excavate

Exercise Three

1. 我们开车去上班或上学所行使的道路，所经过的桥梁以及我们工作所在的高楼，这些都是土木工程师设计和建造的。

2. 因为包含范围太广，土木工程学又被细分为大量的技术专业。

3. 他们也可能管理规模从几个到百个雇员的私营工程公司。

4. 土木工程还包含了其他学科，例如海岸工程、建设工程、地震工程、材料工程、交通工程以及测绘学。

5. 土木工程在我们每一个人的生活中都扮演着重要角色，但日常生活中我们可能并没有真正察觉到它的重要性。

Unit 2

Lead-in

Aggregate; concrete; cement

Section A

Exercise 1

1. F 2. F 3. T 4. F 5. T

Exercise 2

1. Civil engineering materials are important for building. There are many civil engineering materials, such as cement, concrete, aggregate and so on.

2. Construction workers usually do some construction work, like the construction of roads, bridges, skyscrapers and so on. And personally speaking, their work is very tiring.

3. Yes, but I only have a general understanding of the process of making cement. /No, I don't have a clear idea about the process of making cement, but I'm quite interested in learning about it.

4. Yes, I have been to the construction site once under the guidance of my teacher. And I gained much practical knowledge. /No, I have never been to the construction site.

5. As far as I'm concerned, both of them are important. We need to learn some theoretical knowledge from textbooks while we are also supposed to learn some practical knowledge from the construction site.

Exercise 3

In terms of purchasing some important civil engineering materials from a foreign company, it is necessary for me to make good preparations. First, I have to know clearly what materials and how many/how much we need. Then I have to discuss with the boss of our construction company about the budget. Then I need to contact the foreign company about our specific purchasing, the names of the materials, the numbers/amount of the materials, the shipping method and so on.

Section B

Exercise 1

1. D 2. C 3. C 4. A 5. B

Exercise 2

1. unite
2. consolidate
3. ground

4. ratio
5. porous
6. compacted
7. discharging
8. adjacent
9. fine
10. durable

Exercise 3

1. 水泥窑通常直径达 12 英尺——大到足以容纳一辆汽车。并且，在许多情况下，水泥窑比一栋 40 层高的建筑物的高度还要长。

2. 冷却器里的热空气回到水泥窑里，这个过程能够节约燃料并增加燃烧率。

3. 通过一个被称作水化的化学反应，水泥浆变硬，硬度增强，形成一种像岩石的物质，这种物质被称为混凝土。

4. 如果混合搅拌物的水泥浆过量，将会很容易浇筑，并制造出一个光滑的混凝土表面；但是，所生产出来的混凝土成本不低，而且很容易破裂。

5. 相对较薄的建筑部分需要小且粗的骨料，而直径达 6 英寸的骨料常被用于建造大型水坝。

Unit 3

Lead-in

bidder; contract; cash deposit

Section A

Exercise 1

1. F 2. F 3. T 4. F 5. T

Exercise 2

1. We expect to open the tender on 1st of November, in Beijing.

2. Yes. You are supposed to pay it on time. If you don't furnish a tender bond on time, your tender will not be considered. If one fails to win the award, the tender bond shall be returned to the bidder within one week after the decision on award is declared.

3. The relevant documents will be sent next month, from which you can find the details.

4. The most important thing is that our company demand the quality be exactly the same as the terms and conditions stipulated in this contract.

5. Please feel assured that we'll do everything we can to ensure delivery.

Exercise 3

L: Good morning, Mr. Johnson. I'm Li Yun, from China Star Construction Company limited.

J: Good morning, Mr. Li. Congratulations! Your company won the tender of the Modern Commercial Building in Dubai.

L: Thank you so much. On behalf of my company, today I am appointed to negotiate some details about our contract.

J: Yes, it's time to do that.

L: Here's the draft contract, Mr. Johnson. Let's discuss the clauses to see if we agree on all of them.

J: Fine. I'd like to go over it first. (After about 15 minutes) Hmm, you've done a pretty good job. It's well prepared.

L: Thank you. Are there any special requirements on the project?

J: In recent years. we've visited a diversity of buildings all over the world, and we hope that our building is the perfect match of advanced technology and attractiveness.

L: We can improve our technology to meet your special need.

J: Can you guarantee delivery in time ?

L: Please feel assured that we'll do everything we can to ensure delivery.

J: Yes. Very good. Isn't time to sign the contract?

L: Here are two of the originals of the contract, please countersign them.

J: Done, hope we have a nice cooperation.

L: I think so.

Section B

Exercise 1

1. B   2. D   3. B   4. D   5. C

Exercise 2

1. voluntariness
2. engineering
3. facade
4. drainage
5. observe
6. stipulated

7. submit

8. Arbitration

9. signing

10. prevail

Exercise 3

1. 本合同由如上列明的，甲、乙双方按照《中华人民共和国合同法》遵循平等、自愿、公平和诚实信用的原则签订。

2. 如甲方认为乙方确已无能力继续履行合同的，则甲方有权解除合同，乙方必须在接到甲方书面通知后两周内撤离场地。

3. 乙方应参加甲方组织的施工图纸或做法说明的现场交底，拟订施工方案和进度计划，交由甲方审定。

4. 乙方应每月按时向每位施工人员发放工资，如果乙方未按规定发放工资而给甲方带来的任何损失，则乙方承担完全赔偿责任。

5. 如因不可抗力，诸如火灾、水灾、政府强令措施以及其他不可抗力的原因致使乙方不能按期完成项目，乙方不负违约责任。

Unit 4

Lead-in

construction project; negotiation; construction crew

Section A

Exercise 1

1. T   2. F   3. F   4. T   5. T

Exercise 2

1. Within two weeks.

2. That is due to recent adverse weather conditions. We are trying our best to make up for the lost time.

3. Because of a breakdown in the crane, we will probably be a few days behind with the completion of the erection bay.

4. I think we have the right to do so when your payment to us is delayed. May I refer you to Clause 18 of the Contract?

5. Don't worry. It won't be affected and a lot of money will be saved if the design is modified a little.

Exercise 3

S: Mr. Ma, would you please give us an account of what you have achieved for the past month?

M: Ok. Let me make a brief introduction of progress. Generally, the cofferdam, and the tunnel excavation will be finished within the next few days and on schedule. For the power house, because of a breakdown in the crane, we will probably be a few days behind with the completion of the erection bay.

S: Thank you, Mr. Ma. From Mr. Ma's report, we can see that the progress, in general, conforms to the construction schedule with which we are quite satisfied.

M: Thank you. Good cooperation between us will certainly enhance our success with this project.

S: I can assure you, Mr. Ma, as the representative of the project owner, I will, as always, stand firmly behind you in the execution of this project. And I believe that, through our mutual efforts, this project will surely be a success.

Section B

Exercise 1

1. A   2. D   3. B   4. A   5. B

Exercise 2

1. Temporary
2. Ambitious
3. Manual
4. Execution
5. Budget
6. Integrated
7. Modification
8. Available
9. Accumulative
10. deficiencies

Exercise 3

1. 一个项目有明确的起止时间，需要在一定时期内为之努力，需合理安排以生产出一个独特的产品、实现一种特殊的服务或完成一个特定的任务。

2. 项目的临时性与普通商务（或业务）形成对比，后者是重复的、永久的或半永久的生产产品或服务的功能性活动。

3. 执行过程组要确保项目的可交付成果按管理计划生产执行。

4. 未能充分计划会大大降低项目成功实现其目标的机会。

5. 如果这个阶段执行不好，项目不可能成功地满足业务需求。

Unit 5

Lead-in

Landslide; piling; deep foundation

Section A

Exercise one

1. T　2. F　3. T　4. T　5. F

Exercise Two

1. First, we make the site even and solid enough for freely piling machines.

2. We find out and take away the barriers on the piling areas.

3. More than twenty years.

4. We'll stop driving and study the driving records, and then make holes to find out the real situation under earth.

5. To keep the quality of the underground works complied with the specification.

Exercise Three

W: Good morning! Shall I visit the piling site?

L: Sure. Follow me please.

W: I'd like to ask you some questions about piling, Mr. Li.

L: Never mind.

W: What do you do before piling?

L: We make the site even and solid enough for freely moving piling machines.

W: While piling, how do you solve some problems?

L: If the pile can't be driven further deep but the depth of poling is still far from the design, we'll stop driving and study the driving records, and then make holes to find out the real situation under earth, and then we should consult with the designer.

W: What will you do if the barrier is found?

L: We'll make a decision if we change the position of the pile or not.

W: Once you change the position of the pile, what must you pay attention to then?

L: Anyway, we must keep the piles reaching the depth and loading capacity without making any damage to them.

W: Speak on the grounds. It's clear from the way you talked that you know the subject quite well.

L: It's self-evident. If necessary, we'll consult with the designer again until we're a hundred percent certain we can succeed in piling.

W: Good. That's exactly what I want to know. Thanks a lot.

L: You're welcome.

Section B

Exercise one

1. D   2. B   3. A   4. C   5. D

Exercise two

1. transmitted

2. geotechnical

3. elasticity

4. Voided

5. alternate

6. stiff

7. mechanics

8. predominance

9. disturbed

10. cantilevered

Exercise three

1. 岩土工程师确定和设计将要建造的人造结构所需的基础，土方工程和路面等级的类型。

2. 基础是建在地面以下用来传递竖向荷载的混凝土基座。

3. 在松散土层中开挖需要下套管。

4. 基础隔离技术的基本概念是再结构框架的基础中引入水平方向的柔性。

5. 需要对工程坡度设计和预估自然坡或设计坡度失稳的风险进行稳定性分析。

Unit 6

Lead-in

freeway; maintenance; construction

Section A

Exercise 1

1. T   2. F   3. F   4. T   5. F

Exercise 2

1. The excavators, bulldozers and scrapers are needed for excavation and pavement.

2. Because of the bad weather, the delivery of building materials is delayed.

3. A two-shift system will be applied to guarantee the smooth progress of the project.

4. Mr. Zhang is the project manager from China Construction Eighth Engineering Division Company Limited.

5. For example, various facilities have been constructed for special treatment of waste water so as to prevent contamination.

Exercise 3

Student A=A   **Student B=B**

A: How is everything with you?

B: So far so good.

A: Well, the deadline of the Nanchong-Chengdu Highway Project is approaching in accordance with the contract. Is that right?

B: Yes, the deadline is at the end of this month.

A: How is the project going?

B: Generally speaking, the case as a whole is not bad. However, it is reported that a typhoon is on its way. The construction may be stopped for a week.

A: Will the project be completed on schedule?

B: Don't worry about it. We will apply a two-shift system to guarantee the project progress.

A: That would be fine! It is necessary for you to make sure that the drainage system works properly. More importantly, attention should specially be drawn to safety issues at the construction site.

B: Some workers have been designated to be in charge of this job.

A: Sounds great!

Section B

Exercise 1

1. A 2. B 3. C 4. D 5. D

Exercise 2

1. adjacent

2. removed

3. eliminations

4. asphalt

5. prolong

6. pavement

7. flexibility

8. excavated

9. drainage

10. maintenance

Exercise 3

1. 在中国，现代公路建设促进了国民经济的快速发展和人们生活水平的提高。

2. 现代社会力求公路建设对环境和社会影响的最小化，（人们花费）更多的精力和时间进行公路设计。

3. 精良的设计方法、先进的分析技术和电脑辅助制图的应用已经整体提高了公路建设的质量。

4. 当交通、财政和环境方面的注意事项处理完成后，测量员开始对公路进行放线。

5. 路面类型的经济可行性在很大程度上取决于建筑材料的成本和实用性。

Unit 7

Lead in

Lead-in

rigid beam; suspension; arch

Section A Exercise 1

Exercise One

1. T 2. F 3. T 4. F 5. T

Exercise Two

1. Ten years. Here is a letter of recommendation from my previous employer.

2. The company I worked with is too far from my home. Besides, the salary is not enough for me to support my family.

3. The first step is to check the engine oil, hydraulic oil, brake fluid and compressor oil. You must be certain that the oils are within the specified ranges of the oil level gauges. To check the oil levels accurately, you need to place the machine horizontally. If you want to recheck the oil level after filling with oil, wait for several minutes. It takes some time for oil to reach the sump. Secondly, you have to see whether there are any loose bolts or damaged oil hoses. Tighten the check nuts if necessary. After you finish all this, you can start the engine.

4. Next Monday.

5. My ideal pay is 1,000 dollars per week.

Exercise Three

J = Job Seeker   Z = Zhang

J: Excuse me, I saw in your advertisement that you have some vacancies for dump truck drivers. Are they still available?

Z: We still have two vacancies for ten-wheel dump truck drivers.

J: Lucky enough. I am looking for a job as a driver.

Z: May I see your driving license?

J: Certainly, here it is.

Z: How long have you been driving this kind of truck?

J: Ten years. Here is a letter of recommendation from my previous employer.

Z: Fine. May I ask why you quitted the job there?

J: The company I worked with is too far from my home. Besides, the transportation is not convenient.

Z: Do you have your ID with you?

Job Seeker: Yes. Here you are.

Z: And your Social Security

J: I had one, but lost it last week. If you accept me, I will apply to the Social Security Board for a new one. I think I can get it within a week.

Z: OK. This is an application form. Please fill it in. Tomorrow you come here for a driving test. If you pass it, you will be a probationary worker for two weeks. If you prove to be a qualified driver during that period, you will become a formal worker and can work with us until this bridge project is completed.

J: How much do you pay, please?

Z: The starting wage is 25 dollars for a day shift and 30 for a night shift. One shift is seven hours and a half, with extra pay for overtime work, if any. You will have the chance to get a rise if you prove to be a good worker.

J: Do I have to live on the site camp? If not, how do I get to the job site everyday?

Z: We provide free transportation for all our workers.

J: That's good. So what time do I come here for the test tomorrow?

Z: Between 10:00 and 11:00 in the morning.

J: Thank you, sir, I'll see you tomorrow.

Z: OK, see you tomorrow and good luck.

Section B

Exercise One

1. D 2. B 3. B 4. C 5. A

Exercise Two

1. Spans
2. Basically
3. Bridging
4. Used
5. Tension
6. Continuous
7. Closure
8. Anchored
9. Suspending
10. redirected

Exercise Three

1. 到目前为止，大部分桥梁都是公路桥或铁路桥，也有少部分人行天桥。

2. 在河面宽度超过单根树木长度的地方，中国人和印度人用两堆树干从两岸向河心而建。

3. 这种结构叫作突拱，看起来是阶梯式而不是拱面平滑的砌石拱。

4. 通过桥墩连拱便可跨越小河，且外观优雅别致，质量过硬，先前的任何结构形式均无可匹敌。

5. 悬索桥最简单的例子是杂技演员高空走钢丝用的钢丝，原始的悬索桥不过是很少的几根这种钢丝系在一起，供手扶和站立。

Unit 8

Lead-in

rolling stock; maglev; HSR

Section A

Exercise One

1. T 2. F 3. F 4. T 5. T

Exercise Two

1. About tow and a half miles

2. It should be completed by December next year according to the Contract. if everything goes smoothly, it will be complete 2 month ahead of time.

3. There is no standard width for railways around the world. For China, it's 1.44 meters, it's wider, so theoretically they can have higher speed trains and can carry more freight. In many Southeast Asia countries, they have the one-meter gauge.

4. From Chinese perspective, getting access into Laos is very important because they have a number of mining concessions, and linking with Thailand gives them another access to the sea. For Thailand, it is a good opportunity to stimulate economy.

5. "ASEAN" refers to Association of Southeast Asian Nations 东南亚国家联盟

Exercise Three

Hi everyone, I am very honored to give this report to all of you on behave of the discussion group. As for this question, each of us contributed his or her opinions respectively. In general, we believe that "rolling stock" is about train cars. The term rolling stock originally referred to any vehicles that move on a railway. It has since expanded to include the wheeled vehicles used by businesses on roadways. It usually includes both powered and unpowered vehicles. Today there is a very wide range of rolling stock, which used throughout the world on different railways. This range includes the following basic types: Locomotives; Freight wagons; Passenger coaches; Multiple units (with motive power in-built); Metro cars; Light rail/Trams; Rail mounted machines; Inspection and maintenance trolleys; and so on. That's all. Thanks.

Section B

Exercise One

1. A 2. B 3. D 4. B 5. D

Exercise Two

1. consists
2. intercity
3. upgraded
4. Notable
5. rebounded
6. commercial
7. operated
8. suggested
9. rounds
10. promoted

Exercise Three

1. 高铁线路包括新建的客运专线铁路（客运专线）和通过改建的、可同时容纳乘客和货物的城际线路。

2. 中国高铁自 2007 年 4 月 18 日正式投入运行以来受到了广泛的欢迎，2016 年高铁的年载客量超过了 14.4 亿人次，中国高铁也因此成为世界上最繁忙的路线。

3. 这辆蒸汽动力的火车从满洲里南部的大连驶往新京（长春），最高运营速度 110 千米/小时，最高测试速度 130 千米每小时。

4. 1993 年，中国商务铁路运输平均时速为 48 千米，与航空运输和不断扩张的高速公路网络运输相比，渐渐丢失了市场份额。

5. 2007 年 4 月，中国铁路第六轮"提速"圆满完成，多条线路实现了高铁运输，时速达 250 千米的线路累计 423 千米，时速达 200 千米的线路累计 3 002 千米。